Vexing Nature?

On the Ethical Case Against Agricultural Biotechnology

Vexing Nature?

On the Ethical Case Against Agricultural Biotechnology

by

Gary L. Comstock
Iowa State University

KLUWER ACADEMIC PUBLISHERS
Boston / Dordrecht / London

Distributors for North, Central and South America:
Kluwer Academic Publishers
101 Philip Drive
Assinippi Park
Norwell, Massachusetts 02061 USA
Telephone (781) 871-6600
Fax (781) 681-9045
E-Mail <kluwer@wkap.com>

Distributors for all other countries:
Kluwer Academic Publishers Group
Distribution Centre
Post Office Box 322
3300 AH Dordrecht, THE NETHERLANDS
Telephone 31 78 6392 392
Fax 31 78 6546 474
E-Mail <services@wkap.nl>

Electronic Services <http://www.wkap.nl>

Library of Congress Cataloging-in-Publication Data

Comstock, Gary, 1954-
Vexing nature? : on the ethical case against agricultural biotechnology / Gary L. Comstock.
p. cm.
Includes bibliographical references.
ISBN 0-7923-7987-X (alk. Paper)
1.Agricultural biotechnology—Moral and ethical aspects. I.Title.

S494.5.B563 C653 2000
631.5'233—dc21

00-064014

Printed on acid-free paper. Printed in the United States of America

The Publisher offers discounts on this book for course use and bulk purchases. For further information, send email to <molly.taylor@wkap.com> .

for Lisey,

pretty-goodest

Contents

Acknowledgements

I venture, at the risk of overlooking someone, to acknowledge those who helped to improve this manuscript. My deepest gratitude, then, to:

Rich Noland, who argued with me about every idea in the book. All factual errors and mistakes in reasoning should be attributed to him.

Ned Hettinger, who responded with dozens of pages of single-spaced comments on drafts of nearly all the chapters. Would that I had answers for each of his questions.

Tom Regan and Gary Varner, whose writings changed my life. Paul Thompson, whose work in agricultural ethics pioneered the field.

Robert McKim and Jeff McMahan at the University of Illinois; Fred Gifford and Tom Tomlinson at Michigan State University; Lilly Russow and Martin Curd at Purdue University; Tom Regan and Tim Hinton at North Carolina State University; Courtney Campbell at Oregon State University; and Humberto Rosa at the University of Lisbon, all scholars with whom I have had the privilege of planning various ISU Model Bioethics Institutes, and with whom I have discussed many of the arguments in these pages.

Hector Avalos, Bob Hollinger, Margaret Holmgren, David Kline, Joe Kupfer, Bill Robinson, David Roochnik, Tony Smith, Michele Svatos, and Dik VanIten, collegial commentators on various drafts.

Kathryn P. George, whose review of a book I edited on family farms prompted me to write "The Rights of Animals and Family Farmers," now part of chapter 3, one of my first attempts to articulate my changing views about the moral status of animals. David Detmer, whose reply to my paper "Should We Genetically Engineer Hogs?" helped to refine ideas that have made their way into chapter 3.

Emery Castle and Bruce Weber, who made it possible for me to write chapter 2 by inviting me to be a Center Associate of the National Rural Studies Committee at the Western Rural Development Center, Oregon State University, 1989. Peter List and Kathleen Moore, who extended several courtesy summer appointments to me as a member of the OSU Philosophy Department.

Art Caplan, Bill Frey, Glenn McGee, Mike Meyer, Phil Quinn, James Serpell, Tom Shanks, Bill Spohn, Max Rothschild, and colleagues in the ISU Department of Animal Science, for comments on the intrinsic objection to plant and animal genetic engineering, now incorporated into chapter 5.

Will Aiken, Robin Attfield, Bryony Bonning, Baruch Brody, Allen Buchanan, Douglas Buege, Steven Burke, Fred Buttel, Charlotte Bronson,

Baird Callicott, Marti Crouch, Don Duvick, John Fagan, Walt Fehr, Clark Ford, Harry Frankfurt, Jack Girton, Rebecca Goldburg, Richard Haynes, Thomas Imhoff, Wes Jackson, Dale Jamieson, Hugh Lehman, John Mayfield, Mardi Mellon, Tom Peterson, C. S. Prakash, Bernie Rollin, Max Rothschild, Mark Sagoff, David Schmidtz, Steve Shafer, Steve Sapontzis, Loren Tauer, Paul W. Taylor, Luther Tweeten, Vivian Weil, Bob Zimbelman, and Bob Zimdahl.

Jack Dekker, Steve Radosevich, Homer LeBaron, and Richard S. Fawcett, who taught me what I know about the science and economics of herbicide resistance and saved me from major errors in chapter 2.

Lisa Peters, Dan Nebbe, Shileen Groth, Lisa Kane, Aimee Houser, Brad Perri, and Lynette Edsall, Program Assistants for the Iowa State University (ISU) Bioethics Program, who typed much of the manuscript and contributed valuable editorial remarks as well. Cyan Pharr, aspiring veterinarian and promising animal ethicist, for composing much of the Index as part of her Honors project.

Edna Wiser and Janet Krengel, incomparable secretaries and good friends.

Molly Taylor, editor at Kluwer.

Krista, daughter of verve and verb, who used her editorial skills to make changes on a late summer version of the manuscript in Manzanita, Oregon.

I appreciate the institutions that support me. Iowa State University provides me with a fifteen-foot-ceiling corner office on the fourth floor of the refurbished 1892 Agricultural Hall, once Old Botany, now Carrie Chapman Catt Hall. Here, surrounded by hundreds of books on oak shelves, I look down on mature gingko and eastern white pines rimming the playground of the university's new lab school. When I complain to my relatives that writing is hard work, I get little sympathy. Nonetheless, absent a quiet place to work and generous financial assistance, I would never have completed this manuscript.

My thanks, therefore, to these ISU units: the Graduate College, Liberal Arts and Sciences College, Office of Biotechnology, and Bioethics Program. Additional supporters of my research include the National Science Foundation (NSF); the Center for Theological Inquiry, Princeton, New Jersey; the US Department of Agriculture; and the National Agricultural Biotechnology Council, at Cornell University.

Willamette University permitted me to work in the Mark Hatfield library on the glorious intermittent occasions I have had to live and write in Salem, Oregon, where Doug McGaughey graciously allowed me the use of his office and computer. Wheaton College, dear alma mater, granted me access to a word processor and printer one hot Illinois summer.

I mentioned my good fortune in being associated with the ISU Model Bioethics Institute. In helping to conduct these week-long faculty development workshops, I have been lucky to learn some biology from hundreds of outstanding life scientists who come to the Institute dedicated to incorporating discussions of ethical issues into their classes. The Bioethics Institute would not have been possible, however, without the support of Rachelle Hollander, Director of the Ethics and Values Studies Program at NSF.

I am particularly indebted to individuals at ISU who helped to release me from other responsibilities so that I could pursue scholarship: Patricia B. Swan, former Vice Provost for Research and Dean of the Graduate College, who originated the idea of the ISU Bioethics Institute. Walter Fehr, Director of the Biotechnology Program. David Kline, William S. Robinson, and Michael Bishop, successive chairs of the Philosophy and Religious Studies Department.

I dedicate the book to my beautiful Norwegian beloved: quilt designer, teacher, and author; actor, administrator, punster; mother and wife extraordinaire. Seemingly never vexed, she makes us laugh.

Jeg elsker deg.

Introduction

Agricultural biotechnology refers to a diverse set of industrial techniques used to produce genetically modified foods. *Genetically modified* (GM) foods are foods manipulated at the molecular level to enhance their value to farmers and consumers. This book is a collection of essays on the ethical dimensions of ag biotech. The essays were written over a dozen years, beginning in 1988.

When I began to reflect on the subject, ag biotech was an exotic, untested, technology. Today, in the first year of the millenium, the vast majority of consumers in the United States have taken a bite of the apple. Milk produced by cows injected with a GM protein called recombinant bovine growth hormone (bGH), is found, unlabelled, on grocery shelves throughout the US. In 1999, half of the soybeans and cotton harvested in the US were GM varieties. Billions of dollars of public and private monies are being invested annually in biotech research, and commercial sales now reach into the tens of billions of dollars each year.[1] Whereas ag biotech once promised to change American agriculture, it now is in the process of doing so.

The ethical issues associated with ag biotech are diverse and complex. Many worry that genetic engineering might produce unanticipated allergens in previously safe foods; or unexpectedly toxic health supplements; or novel GM diseases. Or environmental catastrophe. Or bizarre new lines of animals possessing genes taken from humans. Or exceedingly wealthy corporations more powerful than the nations trying to regulate them. Or bankrupted family farmers in the US and Europe. Or exploited peasant farmers in developing countries. Or inhumanely treated animals in our labs and on our farms. Or corrupted attitudes to nature among our children.

The book begins with one of the first articles to oppose ag biotech on explicitly philosophical grounds, "The Case Against bGH." Also known as bovine somatotropin (or, BST) and recombinant BST (rBST), bGH is a serum containing a genetically modified protein that farmers inject into dairy cattle to increase milk production by as much as fifteen percent. The first ag biotech product to hit the market, bGH seemed to me in the late 1980s as the most suspect of the early GM products. It seemed destined to single-handedly bankrupt large numbers of family dairy farmers and indirectly to cause various other disruptions in the social fabric of rural communities. Believing that we should be saving small and medium sized farmers rather than driving them out of business, I argued that bGH was a premature technology foisted onto the public by the well-heeled advertising departments of a handful of multinational corporations.

Ethical concerns: Family farms

When I began to write, stories were appearing regularly in Iowa's main newspaper, *The Des Moines Register,* about depressed farmers, stress in rural areas, and suicide. Farm ledger sheets showed high debt loads and low profit margins, and rural businesses faced record rates of foreclosure. It was a time labeled by the media as "the farm crisis," and families throughout the region were palpably strained by economic pressures.

Believing that moral philosophers should address the concerns of those around them I, an assistant professor at Iowa State University, edited a book on ethical issues involved in the farm crisis. The book appeared in 1987 and was titled *Is There a Moral Obligation to Save the Family Farm?*[2] The conclusion to the book argues that there is no direct moral obligation to "save" any particular family farmer, but there are good reasons to try to preserve our system of medium-sized, owner-operated, farms. The system represents a politically and economically viable structure by which we can meet our social obligations to distribute resources equitably, to treat animals humanely, to care for land properly, to nurture mature citizens, and to sustain vibrant rural communities. Whereas there is no direct duty to save this or that farm, a strong case can be made that there is an indirect duty to pursue policies likely to have the effect of saving something like the present system of family farms.

When I had finished writing, my brother-in-law Rich, ever the skeptic, quizzed me about my picture of the ideal farm.

"Just what would a farm look like, if it were *philosophically* and *morally* justifiable?" he asked, suppressing a cynical grin.

Unable to resist a good question no matter how impertinently put, I began to think more broadly. What larger vision ought to guide us in shaping agricultural policy? And how ought we to regard ag biotech, if we want farming to be "morally justifiable?" The questions are not easy, but I came to believe that they have a relatively simple answer, once we find the right place to begin.

The right place to begin is with the question, What is a good farm? and the answer comes from one of America's most prophetic writers. A good farm, argues Wendell Berry, is a farm that does not destroy either farmland or farm people.[3] It is one thing to farm merely to earn cash income, another to farm well. To farm *well*, in accordance with the appropriate standards of excellence, is to produce food and fiber without harming the land or its inhabitants. But just what would it mean to farm in this way? and, What would be the implications of this dictum for agricultural biotechnology?

Neither a farmer nor a political scientist, I had no good answers when my brother-in-law wanted more details. I referred him to Berry's writings and the writings of Wes Jackson, Marty Strange, Gene Logsdon, and Donald Worster, who outline policies they believe would help us to instantiate Berry's vision.[4] Worster, for example, defines good farming as farming that makes people healthier, that promotes a more just society, and that preserves the earth and its network of life. Strange offers specific policy recommendations, suggesting revisions to the tax code and inheritance laws. But in the end, I knew that my brother-in-law, a business executive and former economics professor, could figure out the policy implications on his own. Meanwhile, I had become fascinated by the new biotechnologies being developed for agriculture. I wondered whether the greatest obstacle to Berry's ideal might come from the development of these tools.

In the 1940s and 50s, tractors, fertilizers, and high yielding seed varieties transformed agriculture in the United States. Now, the technologies of genetic engineering were on the verge of dramatically reshaping it again. Would the vaunted new products of ag biotech help to reform agriculture in the direction of Berry's ideal? It hardly seemed so.

Genetically modified plants

Since bGH seems to put dairy cows under additional physiological stresses, I began to consider a broader question, How ought we to treat farm animals? I made a note that I needed to attend more carefully to the issue of animal welfare and rights, but an appointment to the National Rural Studies Committee in 1989 gave me the opportunity to immerse myself in the details of one of the first *plant* biotechnologies: genetically engineered herbicide resistant crops (GEHR). GEHR crops are genetically modified organisms (GMOs) with new genes inserted to help crops survive the application of herbicides. Chapter 2, "Against Herbicide Resistance" presents the results of that study, defending a position of "qualified opposition" to GEHR crops.

Genetically modified animals

Returning to the animal issue, I discovered that researchers had successfully injected human genes into a pig in 1985. There are reasons to worry about the insertion of pig genes into humans, but I was interested in the animals. I wondered whether biotech should be constrained by considerations of animal welfare. Berry's vision of good farming requires respect for land along with people, but if we include soils, plants and ecosystems as fit objects of moral concern, should we not also include individual sheep, cows, chickens, and hogs? If we embrace the idea that farmers should adopt sustainable ecological practices and should not destroy their land, should we not also expect farmers to respect the interests of particular animals?[5]

I realized that attributing moral rights to animals was a radical position, and I did not want to defend a view that entails, as animal rights defenders repeatedly point out, that farmers eschew the slaughter of animals. The practice of raising and slaughtering animals is the backbone of the family farm economy.

I faced a personal moral dilemma. I ate meat and defended the farms on which meat animals were raised. But to respond adequately to my brother-in-law's request for a morally defensible vision of farming I felt I needed an answer for those who defend the rights of animals. What is the relative value of animal and human life? I wished increasingly that I could put this question behind me, focus solely on questions about the broader institution of ag biotech, and move on with my customary mores, diet, and agricultural ideals intact. I learned that I could not do so. Chapter 3, "Against Transgenic Animals" explains why.

bGH, herbicide resistant crops, transgenic animals. Three of the best-known products of ag biotech, and I was opposed to all of them.[6] Was there anything virtuous about GM foods?

Global opposition to GMOs

I found myself attracted to those we may call the global critics of ag biotech.[7] These are people who, in addition to opposing individual products of ag biotech, oppose the entire institution.

I was inspired to write against ag biotech not so much by its best-known critic, Jeremy Rifkin, who seems to oppose all technological change, but rather by two thoughtful biologists, Wes Jackson and Martha Crouch. They have arguments one must reckon with, believing high tech gene splicing is the wrong way to try to feed present and future generations and, more than that, that ag biotech is a symptom of a sick society. When we crave silver-bullet technological solutions to complex systemic problems, we are fooling ourselves into thinking that our problems are simple, shallow. Crouch and Jackson are not the only global critics, and Vandana Shiva, Mae-Wan Ho, Margaret Mellon, John Fagan, Michael W. Fox, Jack Kloppenburg,

Jack Doyle, and the investigative writers at the Rural Advancement Foundation International have all argued in complementary ways. They want to correct the underlying general causes of chronic hunger and environmental degradation before we look for overly simple technological solutions.[8]

After thinking carefully about the global case, I slowly drafted the essay that appears here as chapter 4, "Against Ag Biotech." As the article suggests, I was very nearly a true believer. Rather than confining our opposition to this or that specific product of ag biotech, it seemed that we ought to oppose ag biotech itself along with the modern agricultural paradigm it requires. As Crouch puts it in the title of one of her most influential pieces, the very structure of scientific research in agriculture militates against developing products to help the environment, the poor, and the hungry.[9] Or, as Jackson puts it in one of his titles, our vision for the agricultural sciences need not include biotech.[10]

GMOs may help to solve some problems in agriculture, but will it help to solve what Jackson calls the problem *of* agriculture? For Jackson, as for Berry, conventional modern agriculture is an outgrowth of a materialistic culture moving toward a fragmented future in which most people will be utterly alienated from nature. Not thinking this direction the right direction I, somewhat uneasily, joined the global critics and, in my own way, began as active a campaign against the evils of ag biotech as a schoolteacher from Iowa could muster.

Ethics and stories

I was not originally motivated by professional considerations alone to defend family farms. I teach at a land-grant university in Iowa dedicated in part to helping farmers, and I believe that the institution ought to have someone doing research on the ethical dimensions of family farming. But I took up the issue as much out of personal conviction as professional duty. For, while I was not raised on a farm, I come from a long line of Iowa farmers.

My great great grand -father and -mother, J. H. and Elizabeth Brown Pippert, came here from Prussia in the 1860s. At least one of their children, grandchildren, great-grandchildren, or great-great-grandchildren, has farmed ever since in Cerro Gordo county, a short two-hour drive up I-35. The unbroken chain of Pippert farmers in north central Iowa that stretches back more than a century was sorely tested in the mid-1980s when uncle Harold and aunt Sandy faced potential foreclosure on part of their farm. As readers of *Moral Obligation?* are told in its Introduction, my aunt and uncle farm their own 160 acres plus 240 at the old home place. My relatives survived the crisis of the 1980s, and son Jason now rents ground and raises corn and beans alongside his father.

My wife, Karen, an actor from suburban Washington, D. C., and our children, Krista, Ben, and Drew, also grew attached to the land. Our favorite weekend activity a few autumns back was to drive to Nora Springs on Friday night to watch the high school football game. Jason captained the team and starred at fullback while Jenny led cheers from the sideline. We all sipped hot chocolate and talked politics in the stands before stopping at Casey's for pizza. We never said this explicitly, Iowans don't talk this way, but we love grandma and grandpa's place on the north side of the rise in Highway 18 halfway to Mason City. Although Karen and I own none of it, we nonetheless like to consider the land and buildings and communities ours. Or at least, as aunt Sandy reminds, ours on loan from the Creator.

For me, the continued existence of family farms is not an abstract problem in applied ethics. I cannot approach it only with disembodied principles and a utilitarian risk/benefit calculus. I care about the Pippert farm, and like the idea of my relatives making a living on it.

Along with ethical analysis, this book contains stories: Wendell Berry's story about the loss of American culture and agriculture; the Pippert's story about the pressures medium-sized farm families face. The book itself has a narrative structure, presenting my own story about the route by which I came to write these essays. What role should such stories play in discussions of ethics and public policy?

Very little, according to many scholars in the Anglo-American tradition, who typically assume that ethical issues should be treated abstractly. An individual's memories, subjective experiences, attachments to place, desires and dreams should have little weight in the objective work of impartial normative assessment. The defining characteristics of individual persons--their preferences, race, gender, social location--should not interfere with discussions about how various public policies might affect them. According to the reigning orthodoxy, persons are to be treated as generic and interchangeable. Therefore, explorations of ethical issues should be logical and universalistic rather than narrative and particularistic.

Or so contends the reigning orthodoxy. In the last two decades, however, a number of schools of thought have challenged this view, turning instead to what is called the agent's perspective. Communitarian philosophers such as Alasdair MacIntyre, and virtue theorists such as Michael Slote, take care to look at moral issues from the vantage point of particular moral agents. Ethical problems look different from within different narrative traditions, MacIntyre's phrase for the beliefs, values, ideals, hopes, institutions and practices that define communities.[11]

American pragmatists such as Paul Thompson and Eric Katz sound similar themes, insisting on the importance of discussing particular moral issues not in abstraction but with eyes fixed on how they might be resolved.[12] Feminist philosophers such as Carol Gilligan and Karen J. Warren argue powerfully that all persons are different, and that our differences make a difference.[13] And "narrativists" similarly insist on the centrality of stories, virtues, and local resources. Those we might call narrative environmentalists, like Aldo Leopold, have famously shown the importance of a sense of one's geographical place in thinking through these issues.[14]

To approach problems in applied ethics in an abstract, impartial, way, armed only with high-level principles of justice, beneficence, and autonomy, or utilitarian calculations of costs and benefits, is to run the risk of ignoring the rich social contexts in which our problems arise.[15] It is also potentially to overlook the greatest ally we have for solving our problems: the ideals and aspirations of particular individuals. Resolution of moral problems depends on accurate depiction of the range of solutions actually open to moral agents, and without narratives to fill in the complex particulars, the warp and woof of our everyday lives, we cannot expect to see the full range of options available.

As with individuals, so with societies. Studies of social issues are a form of natural history, but they often tell us little about the specific context within which *our* society must solve our problems. Ethicists are learning to develop schematic, simplified, medical case studies into full-bodied historical narratives, and to analyze stories and rhetoric along with arguments and policies.

We should not simply trust our emotions; without reasoned critical analysis, unconsidered intuitions can be dangerous. Some tell stories simply to stimulate passions or solicit obedience, and they do so without doing justice to the other side of the story. We must approach applied ethics empathetically, yes, in close contact with narratives and emotions. But we must also reason rigorously, subjecting the moral implications of our narratives and emotions to philosophical criticism.

Unlike some anti-theorists in ethics, I do not believe that the so-called narrative, or virtue, approach to ethics is antithetical to the so-called principled, or theoretical, approach. Narratives and principles go together, in the classroom as in the courtroom, where jurists pledged to fair-mindedness try to draw casuistical guidance from older, settled, cases. As in law, so in morality; we draw on prior considered judgments which have withstood the test of time to help us reason analogically about difficult new cases. Narrative reasoning is essential in ethics, with equal emphasis on narrative and reasoning.

I come to write about ag biotech from a specific historical and social perspective, and my motives and ideals for writing are relevant. I care about the farms in Cerro Gordo county; worry about my cousins not having the option of continuing in a line of work that has defined the Pippert family for generations; mourn the declining health of the small towns--Hampton, Greene, Rockwell--around Nora Springs. My personal history, emotions, and desires are the background from which my opinions and judgments emerge. And my philosophical opposition to ag biotech springs from the fear that its products will strain my extended family's way of life, a good way of life, honed and perfected over generations of practice and refinement.

I realized early on that there was very little of a practical sort that I could do to help rural Iowa families and communities to survive. I had no experience planting corn; no spare time during fall semester to help with harvest; no deep pockets. I resolved to do what I knew how to do, write essays, hoping thereby to lend support to public policies designed to help families stay on their farms. Given the strength of my emotions and my belief that ag biotech would vex nature--farmers, animals, ecosystems--I saw my attempts to present the ugly side of ag biotech as acts of resistance.

Changing stories

I floated the global argument in oral presentations to students in my classrooms at Iowa State, in lectures to various audiences at other universities, and even to the occasional radio interviewer. However, I never felt comfortable putting it into print. Will all GMOs be unsafe? Will none of them move us toward our best ideals of farming? Will ag biotech inevitably lead to fewer and larger farms? Should all GM foods be rejected? As I worked on successive drafts of what appears here as "Against Ag Biotech," I thought I knew the answer. Yes, we should oppose ag biotech unconditionally, and that is the answer I defend in the essay published in chapter 4. That answer, however, is not my final word on the subject. I ask the reader to approach chapter 4 as an historical rather than definitive document, the momentary culmination of a certain thread of my thoughts.

In the years since developing my version of the global argument, I have continued to mull over the literature in agricultural economics, ethical theory, political philosophy, ecology, agroecology, environmental ethics, animal rights, entomology, microbiology, weed science, and ethology. Alas, my views have evolved. I no longer believe in the global case.

Why have I changed my mind? The reasons are complex. Briefly, they have to do with my sense of resignation before the fact that government interventions in the market probably cannot save family farms. That genetically modified crops probably are and will be safer for people and the environment than the current crops and pesticides now in use. That nature probably is not properly construed as an individual and, therefore, not a subject with the capacity to be helped or harmed, benefitted or vexed. That many animals lack feelings and a future and, therefore, are eligible for transgenic experimentation, all other things being equal. That some ag biotech products may indeed help us to meet obligations we have to assist those less fortunate than ourselves.

The last two chapters explain these reasons in some detail. Chapters 5 and 6 outline the arguments that led me, an early and somewhat vocal critic of ag biotech, to change horses. I have come to believe, on ethical grounds, that we ought to endorse many GM crops and foods.

Notes

[1]. Total biotechnology sales in the US in 1995 came to $6.8 billion. Healthcare biotech products accounted for approximately two-thirds of this figure, with ag biotech accounting for approximately 20 percent of it. “Explosion in biotechnology,” Agri-food Biotechnology without borders, First International Convention of Science parks Specialized in Agri-Food and Biotechnology, Quebec, April 1996.
[2]. Gary Comstock, ed., *Is There a Moral Obligation to Save the Family Farm?* (Ames: Iowa State University Press, 1987).
[3]. Wendell Berry, “A Defense of the Family Farm,” in Comstock (1987), p. 348.
[4] Donald Worster, “Good Farming and the Public Good,” in Wes Jackson, Wendell Berry, and Bruce Colman, eds., *Meeting the Expectations of the Land: Essays in Sustainable Agriculture and Stewardship (*San Francisco: North Point Press, 1984), p. 37-40. See also Jackson, *Altars of Unhewn Stone (*San Francisco: North Point Press, 1987); *Becoming Native to this Place (*Washington, DC: Counterpoint Press, 1994); and, with W. Vitek, *Rooted in the Land: Essays on Community and Place (*New Haven: Yale University Press, 1996). Also, Marty Strange, *Family Farming: A New Economic Vision* (Lincoln: University of Nebraska Press, 1989), and Gene Logsdon, *At Nature's Pace: Farming and the American Dream* (New York: Pantheon, 1995).
[5]. Cf. Kathryn P. George, “Defending a Way of Life (A Critical Review of *Is There a Moral Obligation to Save the Family Farm?*) in *Between the Species* 7 (Summer 1991): 148-153, and my response following.
[6] There are various ways to try to answer ethical questions about science and technology. One approach is to focus on massive cultural trends, describing how the human spirit is being shaped, or mis-shaped, by “technology” as a whole. Langdon Winner is one of our more important thinkers who takes such an approach. In *The Whale and the Reactor: A Search for Limits in an Age of High Technology* (Chicago: University of Chicago Press, 1986), Winner treats a huge subject, constructing his position out of observations about technologies as diverse as manned space travel, nuclear reactor power plants, and mechanical tomato harvesters. Cf. his first book, *Autonomous Technology: Technics-Out-of-Control as a Theme in Political Thought* (Cambridge: MIT Press, 1977).

Another approach works more from the bottom up, as it were, focusing in detail on particular technologies in a single industry, concerned less with “technology” and the human spirit than with specific inventions, individuals and communities. Both methods have their virtues. Here, I approach ag biotech from the basement, as it were, analyzing particular technologies. I choose this method not to avoid having to make more sweeping judgments about the direction of culture but rather to let those judgments emerge out of a set of circumscribed observations.
[7] Thanks to David Kline for suggesting this felicitous phrase.
[8]. See, for example, Vandana Shiva, *Staying Alive: Women, Ecology and Development* (London: Zed Books, 1988); Jack Kloppenburg, *First the Seed* (Cambridge, Great Britain: Cambridge University Press, 1988); and Jack Doyle *Altered Harvest: Agriculture, Genetics, and the Fate of the World's Food Supply* (New York: Penguin, 1985). By “the writers at

RAFI" I mean Cary Fowler, Patrick Mooney, and Hope Shand. Cf., the *RAFI Communique* newsletter, e.g., C. Fowler, E. Lachkovics, P. Mooney, and H. Shand, "The Laws of Life: Another Development and the New Biotechnologies," *Development Dialogue* 1988 (1-2): 1-35.

9 Crouch, Martha, "The Very Structure of Scientific Research Mitigates Against Developing Products to Help the Environment, the Poor, and the Hungry," *Journal of Agricultural and Environmental Ethics* 4 (1991): 151-158. As editor of the special issue of *JAEE* in which Professor Crouch's article appears, I take responsibility for not catching the error in the title, which should include the word militates.

10 Wes Jackson, "Our Vision for the Agricultural Sciences Need Not Include Biotechnology," *Journal of Agricultural and Environmental Ethics* 4 (1991): 200-206.

11 Alasdair MacIntyre, *After Virtue: A Study in Moral Theory* (Notre Dame, IN: U. of Notre Dame Press, 1981), chap. 15.

12 Paul Thompson, "Pragmatism and Policy: The Case of Water," in the book edited by Eric Katz and Andrew Light, *Environmental Pragmatism* (London: Routledge, 1996), pp. 187-208.

13 Carol Gilligan, *In a Different Voice*; Karen J. Warren, "The Power and the Promise of Ecological Feminism," *Environmental Ethics* 12 (Summer 1990).

14 Jim Cheney, "Postmodern Environmental Ethics: Ethics as Bioregional Narrative," *Environmental Ethics* 11 (1989): 117-34. A sense of place is central to Aldo Leopold's famous work, *A Sand County Almanac*, and Leopold is often called an environmental ethicist. The designation of Leopold as an ethicist is probably misplaced, as Kristin Shrader-Frechette effectively argued in a paper delivered to the 1999 annual meeting of the International Society of Environmental Ethics. Ethicists analyze arguments, develop moral theories, apply principles to cases. None of these enterprises seemed central to the work Leopold deemed most important for himself.

15 Oliver Sacks, *The Man Who Mistook His Wife for a Hat* (New York: HarperCollins, 1987), pp. viii, 110-111. Also see Martha Nussbaum, *Poetic Justice* (Boston: Beacon Press, 1995).

Chapter 1

The Case Against bGH (1988)

> *There are times when the drive [for technological progress] needs moral encouragement, when hope and daring rather than fear and caution should lead.*
> *Ours is not one of them.*
>
> - Hans Jonas [1]

Bovine growth hormone is a protein that occurs naturally in cattle. A chain of 190 amino acids, bGH is produced by the pituitary gland and helps to regulate a cow's lactational cycle; generally speaking and up to a certain point, the more bGH a cow has, the more milk she gives. Using the techniques of genetic engineering, researchers at Monsanto Company have isolated the gene that produces the protein and devised low-cost techniques to manufacture it. Bacteria are placed into fermentation chambers where they multiply rapidly; lab technicians then extract and purify the final product, which is identical to the naturally occurring protein in 189 of the 190 amino acids; and the product is then injected into cows.

Industry's plan is to sell the product to farmers who will administer it in daily doses to their animals. Monsanto's motivation is not hard to discern: a single dose of bGH may cost them ten cents to make and yet be sold to farmers for fifty cents; a worldwide market of $1 billion a year is predicted by Monsanto's Vice-President, Lee Miller; and a profit ratio of $2 returned for every dollar invested is foreseen.[2] The first agricultural biotechnology to hit the market, it will be commercially available as soon as the Food and Drug Administration finds it safe for consumers. Approval is expected before the end of 1990; the FDA has already concluded that neither natural bGH nor rbGH (recombinant bGH) is biologically active in humans who drink cow's milk.

The product works. Daily injections cause dairy cows to increase production of milk from 10 to 15 percent.[3] And the social benefits seem clear; some farmers will be able to produce more milk from fewer cows using less labor. Dairy operations with large herds are expected to cull their less productive cows, put more feed into the remaining ones, and get the same amount of milk. All this, presumably, while farmers reduce their working hours. As the senior vice president for research and development at Monsanto exclaims, "In the future, a farmer using BST will be able to produce as much milk with 70 or 80 cows as can be produced with 100 cows today, use 15 percent less feed to produce that milk, and finally have a chance to be more profitable!"[4] Consumers are also supposed to benefit; as dairy farmers save money, their decreased costs will be passed along to shoppers in the form of lower milk prices.

With so many benefits promised, why has bGH become anathema to some farm and consumer groups? The farmers' opposition is based on three claims: that bGH is harmful to the environment, constitutes inhumane treatment of cows, and will displace farmers from already-distressed rural communities.[5] Predicting that the use of bGH will drive as many as 30% of all dairy farmers out of business, Jeremy Rifkin has claimed that bGH usage would lead to "the single most devastating economic dislocation in US agricultural history."[6]

I consider the farmers' two claims below.

1. Humane treatment of cows

Several contemporary philosophers have argued that higher mammals such as cows possess all of the characteristics needed to be bearers of moral rights; sentience, purpose, social life, intelligence, emotions, etc.[7] To possess moral rights is just to be entitled to fully equal treatment; we do not countenance discrimination against children with Down syndrome even though they are not as sentient, purposive, or rational as we are. Since they have moral value, they have it fully, and are entitled to equal treatment.

If adult higher mammals possess moral rights, then we must treat them the same way we treat humans who, like animals, lack certain characteristics of normal humans. It is permissible for us to act paternalistically toward them insofar as they need extra care. But we may not exploit those beings who lack a certain measure of linguistic ability or emotional security or physical autonomy. If Tom Regan and Ned Hettinger are right, we ought not to do to cows anything that we would not do to mentally enfeebled human beings; the differences between cows and the "marginal human" cases are morally irrelevant.[8]

On the animal rights view, allowing scientists to administer bGH to cows simply to observe its effects would be similar to allowing scientists to administer it to brain-damaged adults for the same purpose. We would not allow this to be done to any human who was not capable of giving (or withholding) informed consent; consequently, we ought not to allow it to be done to other beings in the identical position.

The strictness of the animal rights view has been criticized as failing to make relevant moral discriminations. For example, moral value is not like a light switch that is either off or on. It comes in gradations, as our ability to acquire more of it (through education) and to lose some of it (by entering an irreversibly comatose state) shows. The quality, intensity, and complexity of different animals' mental and social lives makes them bearers of different gradations of moral value. In addition, it is sometimes appropriate to use an other as a means to our own ends even if the other possesses the full complement of moral value. We do this often, as when we allow attendants to fill our gas tanks, or when we ask our hosts to provide us with a glass of seltzer.

It is not always morally objectionable to use another as a means to our own ends even if that other is the possessor of supreme moral worth. Each of these considerations points to a morally relevant distinction that Regan fails to make in his either/or case (either adult mammals have moral rights in the same sense that humans do or they do not).

A less controversial stance is that animals have gradations of inherent value determined in part by the complexity and intensity of their social and mental life, and that we must act toward them in ways that respect this value.[9] Supposing that we could successfully defend the "humane treatment" of animals view, would the use of bGH be acceptable?

An answer to this question relies on our being able to assess the degree to which bGH-use diminishes the quality of the animals' physical and psychological health and, if it does, whether this harm is justified by the benefits it confers. Accurate data about the long-term effects of bGH are not available, but studies have been completed of the effects of using bGH during one lactational cycle.

bGH works by stimulating the division of muscle and liver cells and, apparently, inhibiting the growth of fat cells. (This is the reason for its attractiveness beyond the dairy industry; beef and swine producers expect it to lead to leaner meat.) Evaluations of the effect of the drug on the overall health of the animal are divided between those who see few if any adverse effects and those who are more skeptical. Don Beitz, animal scientist at Iowa State University, notes that while use of bGH leads to increased feed consumption, bone growth (in young animals), muscle growth (in adults), and milk production, the efficiency of the digestive tract and reproductive system seems to be unaffected; the birth rate of calves is the same for treated and untreated mothers.[10] Beitz acknowledges that treated animals do require more intensive management since their nutrient requirements are greater, but he does not anticipate deleterious effects from proper usage of the protein.

Others are more concerned. bGH will put the cows' body metabolism under greater physiological stress. David Kronfeld of University of Pennsylvania claims that high levels of bGH result in "subclinical hyper-metabolic ketosis, a condition associated with reduced reproductive efficiency, mastitis, decreased immune function and `the full gamut of other diseases typical of early lactation.'"[11] Research at the University of Missouri, according to Kronfeld, also supports the view that the drug negatively effects many animals' reproductive efficiency and health. It is worth pointing out, however, that mastitis--a painful infection of the udder--is a very common problem for dairy cows even without bGH, and that the dangers associated with decreased immune function--lowered resistance to infectious and contagious diseases--may be minimized with good veterinary care.

Both the proponents and critics of bGH are relying on scientific data taken from experiments lasting only a short term. Until we have studies that look at the longer-term effects of bGH, studies covering several lactations, we will not be able to say with much confidence whether the drug seriously impoverishes the lives of the cows or not. But on the basis of what we do know, it seems reasonable to conclude that bGH is relatively safe for the cows if carefully administered: that is, given for one lactational cycle and then in low doses. Under such conditions, the treatment seems no more inhumane than many other practices typical of modern dairy operations.

The objection from humane treatment might lose force if other considerations outweighed it. Do current economic conditions justify the risks associated with bGH usage? If we were at war and milk supplies were endangered, if extreme shortages were anticipated in the short- or long-term, if our children were calcium-deficient because our cows were such poor milkers, then our need to exploit the cows' ability to produce might outweigh the risks to the animals' health. Few would argue, however, that this is the case. In developed countries, there is too much milk, not too little; the United States' Congress is trying to decrease milk production by 8.7 percent by paying producers $1.2 billion to get out of dairy production. Human need for more cow milk does not outweigh the risks associated with the drug's use.

One might argue that bGH is needed in developing countries. Here we would want to look at the broader problems of hunger and poverty in nations such as Guatemala, Ethiopia, and Bangladesh. Do such countries need more milk? In tropical climates, milk production from cows is at a minimum; the weather, for one thing, mitigates against the practice, making the growing of hay and forage, for example, almost impossible. Moreover, many of the people in such cultures would not consume more milk even if it were abundant since they have a natural biological intolerance for it. And finally, infants in these countries ought not to be nourished on cows' milk at all, but on their mothers' milk. So even in the Third World--where one might think that milk production needs a boost--bGH turns out to be a bad answer to an irrelevant question.

We ought also to consider the wider economic dimensions of agriculture in developing countries. Is a capital- and management-intensive technology an appropriate solution to these countries' complex food problems? The style of farming associated with bGH-usage is more adaptable by *latifundios*, large plantation-like farms, than by smaller independent farms. Yet the smaller independently-owned farms hold the most hope for relieving widespread hunger and poverty in the long run. So, even if more milk *were* needed in the Third World, the system of large-scale dairying likely to be required (or induced) by bovine somatotropin is not the answer.

If other considerations justify the risks to dairy cattle associated with intense bGH usage, we have not been shown what they are. The conclusion suggested by this discussion is not one favorable to the marketing and use of bGH. The drug itself is a potential threat to the well-being of the animals as it is likely to be administered to them in doses whose effects are deleterious or unknown. It is also likely to exacerbate the problems involved in the treatment of animals on factory farms. The Wisconsin farmers' first claim--that bGH represents an inhumane method of treating animals--is not without merit for anyone taking seriously the inherent value of animals.[12]

2. Social and Economic Effects

The Wisconsin farmers also called for a boycott against the use of bGH on the grounds that it would dislocate too many producers. The argument here cannot be that the technology will put *some* workers out of business; if we were to object to inventions on those grounds we would have had to oppose railroads, electricity, and electronic printing presses.

Our concerns are raised not when new inventions displace labor, but when new inventions displace labor in ways that seem unnecessary, unfair, arbitrary, or completely unaccompanied by redemptive benefits. People are not infinitely plastic: attachment to place, profession, and way of life is part of human nature. So, even in a market economy in which inventiveness and entrepreneurial independence is valued, it is rational to try to minimize the pain associated with rapid social change, and actively to oppose those changes that benefit only those already most advantaged. Is the new invention needed? If so, how can it be introduced with the least amount of suffering? If not, why is it being promoted and who stands to gain from it? These questions force us to look more carefully at the data about bGH's predicted effects.

Robert Kalter himself has taken pains to point out that his study has been misused by Rifkin. He does not predict that bGH will drive 30 percent of all dairy farmers out of business.[13] He claims that many "technical changes"--including bGH, but not limited to it--combined with the removal of dairy price supports, could cause a 25 to 30 percent increase in the nation's milk supply. Since the demand for milk is relatively static, however, this extra milk would not be consumed. Market equilibrium, then would require a 25 to 30 percent reduction in the number of cows and farms in order to bring supply in line with demand. Since not all farms going out of dairy production would go out of farming, and since bGH is only part of the broader technical change expected in the future dairy industry, Kalter expects that the above scenario might send between 23.3 and 46.0 percent of dairy farmers out of milking.[14]

But this decrease must be compared to what we can expect for dairying without bGH in its future. If the drug is kept off the market, not all dairy farmers will stay in operation; between 17.2 and 20.4 percent of them are expected to go out of business even if there is no technical change. So the technology itself cannot be held responsible for all of the 23 to 36 percent reduction foreseeen by Kalter. How much could be blamed on bGH? If my reckoning is correct, the figures would be between 15.9 and 25.6 percent.[15]

In New York, there were 17,500 dairy farms in 1984. If price supports are removed, Kalter predicts that the number will fall to somewhere between 12,600 and 15,800 over a three year period, depending on the rate of adoption. This decline of 2200 to 4900 is too conservative by the estimates of Magrath and Tauer (1986: 12). They predict that as many as 5400 farms will fail in New York in that period. But they also point out that over the last 10 years, "conventional technological changes and ongoing structural change has resulted in the exit of 4000 dairy farms." Of course, this still means that bGH would take down more dairy operations in three years than had occurred in the last ten years.

We must also put this reduction in the broader history of declining farm numbers. In the years between 1964 and 1984, the United States saw a decrease of 77 percent of dairy farms and, Kalter points out, "this happened without hormone technology."[16] The decrease is due to a number of factors, but the improved efficiency brought about by artificial insemination, embryo transfer and computerized record keeping play a large role. Since the current "farm crisis" has between 9 and 24 percent of all dairy farmers getting out of the business over three year periods, bGH will only add on to the total. This leads Kalter to conclude that bGH will simply "speed up the process a little."

While Kalter's estimates are more conservative than Rifkin's rhetoric, the figures command attention. Technical change (of which bGH will be a part) will be responsible for increasing the expected rate of farmers leaving dairying. Without bGH we can expect at least 17.2 of farmers to go out of business. With it, that figures rises to at least 23.3 percent. Notice, however, that this is an increase of some 33 percent in the number of farm failures. (The number could go as high as 120 percent. Using Kalter's figures for a low inelasticity of demand and a high rate of technical change, farm failures could go from 20.4 to 46.0 percent, an increase of over 100 percent.) If Kalter's numbers seem reliable, then, we might wonder at his judgment. Is a 33 percent increase in the number of dairy farmers forced out of dairying to be interpreted simply as "speeding up the process *a little*?" Is it fair to ask a very small percentage of society to bear all of the costs for a marginal increase in the efficiency of milk production?

Part of the problem with bGH is that it discriminates against small and medium-sized farmers, the same farmers who helped to pay for research on it. The genetic engineering techniques that industry will use in making the protein were perfected at universities like Cornell using public monies. And, in research funded jointly by Monsanto, dairy scientists at that land-grant school tested the validity of the drug while agricultural economists at the same university devised econometric models to gauge its market viability and impact. In both indirect and direct fashion, the potentially displaced farmers paid monies for public research which, in turn, led to private sector developments that promise to put the farmers out of business. Many of these farmers have families that have been in the dairy business for generations. *Prima facie*, then, they are justified in believing they have been treated unfairly.

Assessing the deeper merits of this belief, however, is no simple matter. There are several problems here, touching on issues of fundamental disagreement between social philosophers. What is distributive justice in economic matters? What does it require in this case? Don't the greater benefits brought by the free operation of markets outweigh the social costs incurred in the constant shifting of labor resources in capitalism? If so, isn't bGH really in our common good, even if it displaces one fifth of all our dairy farmers?

Before taking up these questions I want to lay my cards on the table. It is my intuition that the Wisconsin farmers are right; something about bGH's social and economic effects is objectionable. On examination, however, I have found it very difficult to say exactly what that is. No laws have been broken, no contracts circumvented, no federal regulations ignored. Not even Jeremy Rifkin claims that any legal damages have yet been done to any party. So the "injustice"--if we are to call it that--is taking a very strange form. None of it has happened: the 3.2 to 37 percent of dairy failures-due-to-bGH are hypothetical (even if probable) *future* events.

If the oddness of this case tempts us to throw up our hands we will have to resist; if we ever needed a language in which to discuss "potential future injustices" it is now. The skill of social scientists to make sensitive ex-ante studies about the likelihood of various consequences of new technologies grows. As it does, their sophistication in predicting the future quickly surpasses our ability to assess the results of their studies morally. And yet--if it is in our power to do so--it is surely better to prevent an injustice before the fact than to try to remedy one after. So the urgency of trying to assess the farmers' second charge is as great as the conceptual difficulties involved in doing it.

If bGH is unfair to farmers, it is not yet clear how or why. We might begin by specifying the group that, at some future point, is supposed to be the one offended. According to Kalter, bGH is size-neutral; it can be used by farmers whether they have "ten cows or a thousand."[17] Contrary to the claims of bGH's proponents, however, many studies have contested the claim that bGH is size-neutral; the winners and losers will not be evenly distributed throughout the farming population.

Even though bGH may be marketed at a low cost per dosage, successful use of the product will require significant managerial expertise and access to capital. "These constraints," write Barnes and Nowak, "will be most problematic for smaller and less efficient farms that have operators that are less knowledgeable and older."[18] A new technology is not size neutral when its cost-effectiveness improves as the number of cows and the quality of managerial skills increases. And yet, even if individual doses are priced low, larger and younger and better educated farmers will reap disproportionately greater benefits than older, less "aggressive" farmers.

Traditional patterns of technology adoption suggest that larger, more "progressive," producers take earlier advantage of innovations, reaping whatever rewards there might be in increased efficiency.[19] When the rest of the group catches on, these comparative advantages fade. In the case of bGH, early adopters will probably be those dairy farmers with large pedigreed herds, with significant investments in management and labor, access to capital, and low debt-loads. They will be the winners. The losers will be those with high debt-loads or poor soils or small herds or so-called bad management techniques, the producers that the agricultural establishment sometimes calls "inefficient." These are likely to be subsistence farmers in Appalachia, black farmers in the South, and medium-sized farms with high debt throughout the country.

Have the losers been treated unjustly by the agricultural research establishment? An answer to this question requires us to define justice, no easy task. Many definitions have enjoyed favor throughout the centuries of reflection on the matter, but three considerations seem to recur in all of the discussions: equality, contribution, and need. Following contemporary philosophical practice, I will discuss these issues under the headings of distributive justice and the common good.

2.1 Distributive Justice and bGH

The argument from unequal treatment assumes that there is an unwritten contract between agricultural research institutions and the farmers who support them. The farmers pay taxes which go for salaries and equipment; the institutions are supposed to deliver seeds, machines, and techniques that will make farming more productive and profitable for all kinds of farmers. Now, if institutions do research that speaks only to the needs of a certain class of farmers and thereby gives them a comparative advantage over others, then the contract has been broken. The institutions have unfairly privileged one class, and put another at a disadvantage.

There is strong evidence for thinking that smaller and larger-sized farmers have been treated differently. Jim Hightower's book *Hard Tomatoes, Hard Times* popularized the case of the mechanical tomato harvester in California, and the ongoing California court case that resulted from it is adding the weight of legal opinion to Hightower's charge.[20] Of course, some benefits have accrued to small and medium-sized farmers from the university research in question, and these need to be added into the calculus.

Nonetheless, when one considers the kinds of technologies that have come out of agricultural research institutions since the second World War--including, but not limited to, chemical herbicides and pesticides, large tractors and implements, automated milking parlors, artificial insemination, petroleum fueled machines, embryo transfer, and hybrid seeds--a presumption in favor of Hightower's charge appears. Even farmers themselves tend to think that their own farms always need to be a little bigger; there is an ideology of growth in farming that has been caused by, and in turn helps to fuel, institutional research biased toward large-scale, capital-intensive, mechanized agriculture. So the ball is in the opponent's court; the burden of proof is on those who believe that small and medium-sized farmers have *not* been discriminated against.[21]

One might argue that the skewing of research was justified because large farmers assumed a larger share of the tax burden. If the more aggressive operators had paid substantially larger sums, wouldn't they be entitled to the increased attention they received? Even if it were true that big farmers had shouldered most of the burden, this would not justify an unbroken legacy of hard tomatoes and hard tomato harvesters. Which innovations favored smaller producers? Which hybrid seeds, which machines, which chemicals gave assurance that farmers could remain competitive while retaining their present size?

The severity of this research bias would be of one magnitude if small and medium-sized farmers had simply not been able to increase their profits. But the situation is much worse; these farmers have not remained where they were; they have gone through years of financial and emotional upheaval. Many have ended in bankruptcy. As the farm crisis drags on, successive groups of farmers are moved toward the end of a conveyor belt, and dumped over the edge. With each new jerk of the belt, the status quo is changed. No wonder that the US Office of Technology Assessment predicts that medium-sized farms will have completely disappeared by the year 2000.[22]

The extent of the unfairness cannot be seen if one takes a snapshot of the conveyor. The belt is turning, and with each turn, a new group of farmers is dumped off the end. When, as David Braybrooke puts it, "the game begins again," the terms are different. If the results of the last exchange "were unjust, enriching some people at the expense of others, and there are no compensating changes, they bring about a distribution of resources (in private property and in other resources like influence) that raises the prospects of injustice" in the next round of exchanges.[23] As large farmers increase and consolidate their hold on the industry, the universities become even more responsive to their needs, and to the needs of the private sector food processors who prefer to deal with a few large producers. Meanwhile, governmental programs also become increasingly biased toward the larger producers: the amount of governmental assistance provided to large farms increased tenfold between 1980 and 1985, while the assistance given to medium farms increased only fivefold.[24]

The consequences of such unfair exchanges may be even more troubling than the initial injustices. Not only have the medium-sized farmers lost the value of their tax dollars, but they have also given up what Braybrooke calls "increments of power and advantage"[25] that they would have had if the first round had been fair. Their ability to educate themselves about new farming methods, their incentive to organize into effective political units, their skill in bargaining collectively, their capacity to market their goods strategically--all of these skills may suffer serious erosion as a result of the group's having been mistreated in earlier stages.

Whether my theoretical analysis offers a sufficiently nuanced explanation of the history of America's medium-sized farms is arguable. It is admittedly schematic and general. But studies have given us good reasons to believe, more specifically, that 1) prices received by hog and beef farmers in certain portions of the country are artificially lowered because of lack of competition among meat packers in those regions,[26] 2) a concentration in the number of firms in breakfast cereals has artificially inflated prices paid by consumers,[27] 3) that tax laws like rapid depreciation schedules and investment tax credits have favored large producers over small producers,[28] and 4) that the land-grant university system has not taken care to make sure its research is equally beneficial for all sizes of farms.[29]

This list may or may not add up to a longstanding pattern of discrimination by powerful, tax-funded, organizations against the majority of farmers. But the case against bGH does not stand or fall with the answer to that question. Suppose that the process of allocating tax monies for research is judged, as Luther Tweeten argues, *not* to have been biased against family producers. We must still ask ourselves whether the general pattern of the demise of moderately sized farms is socially desirable. In 1986, six percent of all farmers went out of business; one farm every four minutes. In 1985, the figure was five percent. If those figures seem small compared to the general rate of failures of small businesses, consider that most small businesses have only very recently started-up; the farms in question often go back generations. These farms do represent, in the often maligned rhetoric of farm activists, a "way of life" whose value is not measurable in economic terms.[30]

The loss of farmland owned by minorities plays a disproportionately large role in this story. Half a million acres of farmland per year are currently being lost by black owners. The story started, of course, with blacks clearly behind the eight ball; while they constituted approximately 15 percent of the U. S. population, blacks owned almost no farmland at the beginning of the twentieth century. Currently they own 1.4 percent of the farms. Whatever progress black farmers have made, however, is being rapidly eroded. At the current rate, these farmers will be completely landless again by the end of the century.

What is happening to the land? Patterns of land-use vary across the country, but in places where conversion to nonagricultural uses is least problematic, the number of absentee landowners is increasing dramatically. In 1981, the number of acres managed by professional farm management companies was 48 million; in 1986, it was 59 million, an area the size of Colorado.[31] While it is not clear from the data which farms in particular are under the most pressure, it is clear that 66 percent of total farm debt was held in 1986 by medium-sized farms, those usually owned and operated by families who are dependent on them as their major source of income. These are the farms currently closest to the end of the conveyor belt.[32]

What does this story about publicly funded agricultural research and its effects on rural America have to do with bGH? It helps us to see the broader pattern of which bGH is a continuation. If hardships were distributed evenly, if large and small and medium-sized farms--those owned or worked primarily by blacks, whites, and hispanics--had all suffered equally in this tale, then we would have little basis for talking about injustice. But gross discrepancies have been with us for a long time, through several turnings of the belt, and those dumped off the end have not been compensated.

In terms of disparity in income levels and access to power, the situation in agriculture is little different from the wider pattern in the United States. In 1970, the top 20 percent of Americans made 41.6 percent of total family income; the bottom 20 percent made 5.5 percent.[33] By 1985, the top 20 percent were capturing an additional 5.5 percent--up to 47 percent of all earned income--while those on the bottom had dropped to 4.7 percent. Of those working for a living, the most successful in our culture make somewhere in the range of 100 to 200 times the amount of the least successful.[34]

If the discrepancies were temporary abnormalities, we might be able to overlook them. But to the extent that the inequities are deeply entrenched in our history, are likely to persist indefinitely, and are growing worse, they indicate a troubling problem in our agricultural market system. For it is, in Braybrooke's words, "the continual repetition of the discrepancies, with one set of people always faring well, and another always faring badly" that fixes our attention. "Some people, and their children, [are] living their lives out--very possibly shortened lives--without having any chance to live decently; others [are] surfeited with pleasures."[35]

We might defend the agricultural market system by arguing that discrepancies of some magnitude are inevitable in any system of allocating resources, and that the agricultural market system could alleviate gross discrepancies by redistributing resources downward--toward those on the bottom--through political measures such as progressive income taxes. In this case, income transfers (for example, via a truly progressive income tax system) from rich CEOs and agribusiness corporations to seasonally employed migrants and poverty-level farmers would be justified on the grounds of equality and need.

This would be a step in the right direction, but the poor need more than income; they need autonomy, meaningful employment, jobs in which their skills can be used and honed and which help to give them control of their lives. The poor need jobs and education through which to be able to meet their own needs for food, shelter, clothing, and companionship. Farming in the traditional sense has offered that sort of employment. The farmers being put out of business by technological advances do not need income enhancements in the long run. They need secure employment. Thus the answer suggested by Michael Novak--to give farmers cash--may show compassion, but it is not directed toward establishing an agricultural economy that plans rationally and deliberately for just compensation of its members.[36]

It may be objected here that my analysis assumes too much control over the inventive process. How can we *plan* to come up with innovations that would help smaller full-time farmers?

Research on bGH may have begun, in part, because scientists were interested in the molecular structure of a specific protein, but it has been pushed through to the marketing stage only by corporations anticipating significant profits. Expensive biotechnologies do not blossom from people's heads as if they were fresh flowers seeking spring air; they are consciously pursued by powerful organizations with specific plans and needs.

Those who say that "the development of technology" is primarily responsible for the decreasing number of dairy farmers may not intend to mislead us, but they do so when they allow their audiences to infer that history could have followed no other course. In fact, we could have pursued other economic, monetary, and fiscal policies; we could have encouraged farm organizations and cooperatives instead of subsidizing production of targeted crops; we could have concentrated on diversifying our own farms instead of concentrating production on a few export crops; we could have invested in other sorts of research in agriculture--perennial crops, sustainable farming methods, small-scale, non-chemically driven planters and reapers. Those who have the most to gain from large, intensively-managed, petro-chemically dependent methods in farming have played a substantial role in the displacement of farmers.

2.2 bGH and the Difference Principle

How should we go about distributing the benefits of technology? John Rawls suggests that social goods should be distributed fairly, and that inequities in distribution should be accepted only when such inequities will enable those on the bottom to be better off than they would have been if the inequities were disallowed.[37] This is the difference principle: unequal distribution of material goods and social status is fair if and only if it improves the lot of those on the bottom. Poor farmers in the South might be denied certain tax breaks given to bigger farmers if and only if the poorer farmers would come out ahead in the long run. Black farmers might be denied Extension Service attention if and only if this would result in their farm operations improving over the long haul. A progressive tax system would be justified, even though it appears to treat the wealthy unfairly, if and only if it improves the condition of the worst off.

Knowing what we now know about bGH, could we justify denying industry and large farmers profits on the grounds of distributive justice to smaller farmers? Advocates would say no; keeping bGH off the market is unfair to some farmers because it denies them the choice of using it. But, according to the difference principle, this could be justified if it would improve the lives of agriculture's most disadvantaged.

Would a boycott of bGH improve the lot of the worst-off dairy farmers? There are at least two questions here. The first is: Would banning bGH really be good for the marginal farmers? Lester Thurow argues that while there is an excess of farmers, there are plenty of good jobs into which they may move.[38] Rather than artificially trying to save farmers' jobs, society would be better off retraining the farmers, helping them to make the transition into other lines of work. This argument might make sense if we decided not to try to count the psychological costs involved in moving farmers, farm families, and associated rural workers out of their way of life. It might make sense, too, if we looked at the history of farming through deterministic glasses, for if the labor requirements of agriculture have been reduced by inevitable, inexorable, economic forces, then it would be foolish to try to retain workers in farming today. Too many inefficiencies in the allocation of resources are promoted by trying to keep farmers employed.[39]

Laying aside for the moment questions about the validity of this view of history we may still ask whether the argument above takes into account all of the external economic costs involved in moving labor out of agriculture.[40] How much does it cost taxpayers when a displaced farmer moves into an urban area, fails to find a job, goes on unemployment, and eventually loses incentive to look for work? How many tax dollars are spent on Medicare, public nursing, pharmaceutical products, and federal programs in order to care for that farmer? What social costs are incurred by the depopulation of rural areas, the overcrowding of cities, and the malaise and disruption that accompany both?

The fact is that we do not have any idea about the extent of the external costs involved in moving labor out of agriculture. We lack accurate accounting methods "that begin from the assumption that social costs are to be computed so that the public has a far more exact understanding of what any particular item or process costs the society as a whole."[41] So I would not presume to be able confidently to assert that the costs of moving farmers out of their way of life outweighs the benefits of doing so; I have no more privileged way of judging this matter at present than anyone else. What can be asserted, however, is that those who think that they can boldly claim that "retraining farmers" is the only sensible answer to the farm crisis are either naive or privy to divine revelation.

The second question is whether "banning" bGH would be good for the urban poor, many of them grandsons or granddaughters of farmers. A successful boycott against bGH might prevent the lowering of milk prices, or even slightly inflate them and, moreover, have a chilling effect on other avenues of research in industry and university, avenues that might lead to cheaper food for the poor. Advocates of bGH claim that the new biotechnologies will cut costs for farmers, and that these will be passed on to consumers. History, again, is a good antidote for such rhetoric. In recent years farmers have been pressed to cut their input costs while the prices they received on the market for their wheat, corn, and beans dropped steadily. Did the price of corn flakes to consumers drop? During the summer of 1988, many food manufacturers raised prices at the first media stories of the drought. Their costs, of course, had not gone up; they simply used news reports as cover for increasing profits. The facts are that intermediate markets seem to have a way of absorbing whatever profits are made when farmers' prices go down. There is no reason to think that bGH usage would lower milk prices for the urban poor, or any consumers.

2.3 bGH and the Common Good

These considerations compel us to think not simply about distributive justice, but about wider considerations such as the sort of people we are and want to be, the qualities of character we want to encourage in our young, and the type of concerns we wish to pursue together. Our society should be one in which no person goes hungry, in which all who wish to work are employed--in jobs promoting individual autonomy and social cooperation--and in which human flourishing in its moral and spiritual sense is possible. We should pursue objectives that are good, in an objective, substantial, sense; objectives that allow us "to experience the fullness of human life, as opposed to merely existing."[42]

From the perspective of the common good, bGH appears as a technology that not only will fail to promote the common good, but will actively undermine it. It will only add to a decline in the number of dairy farmers, exacerbating the crisis currently affecting rural America. It will degrade rather than enhance the internal goods pursued in the practice of farming since it encourages farmers to treat animals as production machines rather than co-inhabitants.[43] It promises to assimilate dairy farming fully into an impersonal, industrialized culture that farmers have long resisted. In short, bGH threatens to undermine the common good not simply of the dairy farmers it will displace, but of us all. It promises, in a small way, to undermine our general well-being.

That conclusion is worth pondering, and its qualifications worth repeating. bGH *promises* (we should not forget that we are dealing with potential injustices, not yet realized) *in a small way* (it is by no means the most pressing problem in America, though it may be for the less than 1 percent of Americans who are small dairy farmers) *to undermine* (not simply fail to promote) *our general well-being* (it is not simply dairy farmers who are affected, but all of us).

After all of this, defenders of the technology would still have the following response open to them. If we prevent bGH from reaching the marketplace, we may be sending a signal to farmers that inefficient farming is acceptable, and that society will always protect them from innovations that might displace them. This would be counterproductive for society as a whole, making farming a less attractive line of work for farmers, and driving up the cost of food for consumers.

The objection has merit, and it forces us to admit that we walk a fine line when we get in the business of trying to pick and choose between new technologies. We do not want to stifle the imaginative spirit of public or private scientists, nor the independence of farmers for whom farming is attractive precisely because it allows them freedom to try new things. But while bGH is the first agricultural biotechnology, it will not be the last. And discouraging its use in no way commits us to oppose all technologies. We should oppose only those technologies that unfairly advantage one social group over another, that displace workers at unacceptably high costs, or that threaten the stability, beauty, or integrity of the plant or animal kingdom.[44]

3. Policy Recommendations

In the interests of the common good, we ought to pursue at least two goals in agriculture. One is to keep farming open to a wide number of people. The second is to allow innovations that will contribute to the number of meaningful jobs. Accomplishing this goal means matching supply with demand. The free market has not demonstrated the ability to do this in the dairy industry. When left to market forces, dairy farmers--like all farmers--have, in the words of John Kenneth Galbraith,

> a relentless and wholly normal tendency to overproduce, because of extraordinary productivity gains and because farmers, being powerless to influence or control supply and price, harvest more and

more as a way of trying to stay financially afloat.[45]

As Galbraith argues, the answer is a system of supply management, something that is "taken for granted in all large-scale industry." We need a way to organize dairy farmers so that each can make a decent living in a relatively stable business atmosphere without relying on government subsidies or having to try to outproduce one's neighbors. That is a tall political order. My contribution here is only to suggest that the sort of technological direction represented by bGH is of no help in trying to fill it.

4. Conclusion

To the extent that potentially displaced dairy farmers have done nothing for which they ought to be punished; to the extent that the research establishment has clearly favored large producers in its development of techniques and technologies; to the extent that fiscal, monetary and economic policies have disadvantaged small dairy producers; and to the extent that bGH will only exacerbate the unjust consequences of the past; to that extent we ought to oppose this particular biotechnology. Language about "banning" bGH, of course, is just that: a slogan intended to summarize the case against bGH. There is no governmental body with the authority to ban bGH on the grounds of humane treatment of animals. Nor is there any government agency charged with the task of overseeing--much less regulating--technologies by the criteria of their anticipated socio-economic effects. This shows the need for legislative attention to this matter. But in the meantime, opposition to the marketing of bGH sends a signal to those in public and private decision-making positions.

Not all biotechnologies are acceptable. We do not want those that are destabilizing, inhumane, or ugly; we want those that will preserve the beauty, integrity, and diversity of the Creation.

- - - - - - - - - -

Jewish folklore tells of the town of Chelm, whose inhabitants engaged in curious behavior. Knowing full well that rainy season was upon them and that the prayer hall desperately needed a new roof, they spent their time putting new carpet on the floor. The next fall, when their schoolchildren had no papers, pencils, or workbooks, they spent their fortune on another first edition for the rabbi's library. Chelmians always did the opposite of what was in their own interests.

However entertaining fiction may be, contemporary agricultural history is more so. Awash in excess dairy products, our government dumps milk in the ocean, hands out surplus cheese to farmers, and pays operators $1.2 billion to slaughter their cows: all of this while publicly funded institutions quietly spend taxpayers' monies on schemes to increase milk production.

There are daring scientific projects that are in our interests and that need our moral encouragement. bGH is not one of them.

Notes

1. Hans Jonas, *The Imperative of Responsibility: In Search of an Ethics for the Technological Age* (Chicago: University of Chicago Press, 1984), p. 203.
2. Matthew H. Shulman, "Bovine Growth Hormone: Who Wins? Who Loses? What's At Stake?" in Steven M. Gendel, A. David Kline, D. Michael Warren, and Faye Yates, eds., *Agricultural Bioethics: Implications of Agricultural Biotechnology* (Ames, IA: Iowa State University Press, 1990), pp. 111-129.
3. "Experimentally, the greatest response on an annual basis has been a milk increase of 25.6 percent," Kalter, et al., *Biotechnology and the Dairy Industry: Production Costs, Commercial Potential, and the Economic Impact of the Bovine Growth Hormone*. A. E. Research 85-20. Ithaca: Department of Agricultural Economics, Cornell University, 1985: 108. This increase is not expected over the animal's complete lifetime.
4. Howard Schneiderman, "Innovation in Agriculture," *The Bridge* (Spring 1987): 5.
5. Patrick Madden and Paul B. Thompson have addressed the general issue of ethics and agricultural biotechnology in "Ethical Perspectives on Changing Agricultural Technology in the United States," *Notre Dame Journal of Law, Ethics and Public Policy* 3 (Fall 1987): 85-116.
6. Laura Tangley, "Biotechnology on the Farm," *BioScience* 36 (October 1986): 590.
7. Tom Regan, *The Case for Animal Rights* (Berkeley: University of California Press, 1983).
8. See Ned Hettinger, "Cohen on the Use of Animals in Biomedical Research," manuscript, Department of Philosophy, Charleston College, Charleston, SC. Hettinger is responding to Carl Cohen, "The Case for the Use of Animals in Biomedical Research," *New England Journal of Medicine* 315 (1986): 865-870.
9. Among others, Frederick Ferre holds such a view. See "Moderation, Morals, and Meat," *Inquiry* 29 (1986): 391-406. My criticism of Regan is indebted to Ferre's discussion.
10. Don Beitz, "Physiology of Growth Hormone," lecture to Animal Science Roundtable, Iowa State University, May 13, 1988.
11. David S. Kronfeld, quoted in Shulman, 14.
12. I want to reemphasize the promise of biotechnology. If, as is predicted, transgenic animals can be developed that possess "mammary specific genes coding for proteins that enhance the nutritive value of milk or that are biologically important products in their own right," then the milk of healthy, flourishing cows kept in humane conditions could be the source of inexpensive, life-saving, proteins such as insulin or the blood component known as human factor IX.
13. "[Kalter] estimates that the number of dairy farms may have to be reduced 25-30 percent to restore market equilibrium. . . . These adjustments will almost certainly have dramatic social, economic, and cultural effects." Andrew Kimbrell and Jeremy Rifkin, "Biotechnology--A Proposal for Regulatory Reform," *Notre Dame Journal of Law, Ethics, and Public Policy* 3 (Fall 1987): 125. At a conference on "Public Perceptions of Biotechnology" sponsored by the Agricultural Research Service of the USDA at Airlie House in Virginia in 1986, Dr. Kalter objected strongly to a speech by Rifkin in which Rifkin used Kalter's study in the way recounted here.

14. Kalter, 1985, Table 50, “Changes in Price, Output, Employment and Cow Numbers From bGH and a Free Market Policy by Elasticity of Demand,” p. 101.
15. I arrived at these figures by subtracting 20.4 (reduction in farm numbers with 0 technical change and a low elasticity of demand for milk) from 46.0 (reduction with 30 percent technical change) to get 25.6. I then subtracted 17.2 (reduction in farm numbers with 0 technical change and high elasticity of demand) from 33.1 (reduction with 30 percent technical change) to get 15.9 These, of course, are very rough calculations. Again, they refer to the percentage decrease in farm numbers due to technical change of many types, not simply bGH. If farmers started to milk four times a day instead of three, for example, that would lead to a greater output of milk and a corresponding need for adjustment. These qualifications need to be figured into the estimates.
16. Quoted in Tangley, 592.
17. Quoted in Shulman: 20.
18. P. Nowak, J. Kloppenburg, Jr., and R. Barnes, “bGH: A Survey of Wisconsin Dairy Producers,” in *As You Sow* 18 (July 1987), Department of Rural Sociology, University of Wisconsin: Abstract.
19. Fred Buttel, “Agricultural Research and Farm Structural Change: Bovine Growth Hormone and Beyond,” *Agriculture and Human Values* 3 (Fall 1986), p. 98, n. 5.
20. Jim Hightower, *Hard Tomatoes, Hard Times* (Cambridge: Schenkman, 1973).
21. Cf. the confirming opinion of Don Paarlberg, “The Land Grant Colleges and the Structure Issue,” *American Journal of Agricultural Economics* (February 1981): 129-133.
22. Office of Technology Assessment, *Technology, Public Policy, and the Changing Structure of American Agriculture: A Special Report for the 1985 Farm Bill* (Washington, D.C.: U. S. Congress, OTA-F-272, 1985).
23. David Braybrooke, “Justice and Injustice in Business,” in Tom Regan, ed., *Just Business: New Introductory Essays in Business Ethics* (New York: Random House, 1984): 173.
24. “In 1985 . . . the average governmental payment to farms was $5,193 for farms with sales between $40,000 and $99,000 . . . and of $37,499 for farms with sales over $500,000. The same groups of farms received $1,169 and $3,849 respectively, in 1980, which correspond to a fivefold increment . . . and a tenfold increment . . .” Alessandro Bonanno, “Agricultural Policies and the Capitalist State,” *Agriculture and Human Values* 4 (Sp-Sum 1987): 44.
25. Braybrooke: 175.
26. A study by Quail et al. found that if there had been packer competition instead of shared monopoly (four firms) in such regions as Colorado, Nebraska, and Iowa, that the “average price would have been roughly 24 cents per cwt higher and annual returns to feeders in these . . . regions would have been nearly $42 million greater . . .” G. Quail, B. Marion, F. Geithman, and J. Marquardt, “The Impact of Packer Buyer Concentration on Live Cattle Prices.” N. C. Project 117, Working Paper Series, WP-89 (May 1986): 55.
27. In 1972, the Federal Trade Commission attempted to prove this, claiming that the ready-to-eat cereal industry was “highly concentrated (four firm market share of 91) and had high entry barriers. “The Commission was unsuccessful, because “the judge concluded that the defendants had acted like independently behaving rational oligopolists, which is not sufficient to constitute monopolization.” As Bruce Marion points out, this case shows that the courts are not prepared to deal with the complex issue of shared monopolies, choosing instead to apply anti-trust laws (if at all) only to markets clearly dominated by a single company. See Bruce Marion, *The Organization and Performance of the U.S. Food System* (Lexington: D. C. Heath and Co., 1986): 396-397.

28. Cf. Luther Tweeten's view that "Federal income tax provisions have, relatively, most favored part-time small farmers and 'syndicates' financing, for example, large cattle-feeding operations because farm losses can shelter off-farm income and provide large savings per dollar of farm output." He adds that in order to serve social justice we ought "(1) to phase out the investment tax credit and rapid depreciation allowance and (2) to target public program transfers more heavily on farm families with low incomes." Tweeten, (1987): 228, 231.
29. The California Rural Legal Assistance group sued the University of California in 1979, claiming that the Hatch Act of 1887 obligates Experiment Stations to benefit rural constituents, including small farmers, farm workers, and consumers.
30. For an analysis of arguments from emotion, see Comstock, "Conclusion: Moral Arguments for Family Farms," in Comstock (1987): 402-405.
31. The data in this paragraph and the one preceding is from "The Continuing Crisis in Rural America: Fact vs. Fiction," Prairiefire Rural Action, Des Moines, IA, May 15, 1987.
32. Cf. Neil Harl, "The Financial Crisis in the United States," in Comstock (1987): 112-128.
33. Lester Thurow, "Toward A Definition of Economic Justice," *Public Interest* 31 (1973): 77, quoted in Albert Borgmann, *Technology and the Character of Contemporary Life* (Chicago: University of Chicago Press, 1985): 111.
34. Braybrooke, 195.
35. Ibid., 179.
36. Michael Novak, "Cash Income and the Family Farm: Reflections on Catholic Theology and the Democratic Capitalist Political Economy of Agriculture," in Gary Comstock, ed. *Is There a Conspiracy Against Family Farmers? Agricultural Economics, Public Policy and Catholic Theology*, USF Monographs in Religion and Public Policy, Number 5, Department of Religious Studies, University of South Florida, 1989.
37. John Rawls, *A Theory of Justice* (Cambridge: Harvard University Press, 1971).
38. Lester Thurow, "The Agricultural Institutions and Arrangements Under Fire," paper presented to the Social Science Agricultural Agenda Project, Phase 1 Workshop, Minneapolis, MN, June 9-11, 1987: 125-126, citation at 118.
39. Ibid. 124.
40. As Thurow reminds us, if we love "to be or see farmers, it may be rational to protect [our] farmers with tariffs and quotas. What is lost in terms of extra consumption utility is more than gained in extra producer's utility." Thurow, *Dangerous Currents* (New York: Random House, 1983): 121. Quoted in James Montmarquet, "Agrarianism, Wealth, and Economics," *Agriculture and Human Values* 4 (Spring-Summer 1987): 49.
41. Marcus G. Raskin, *The Common Good: Its Politics, Policies, and Philosophy* (New York: Routledge & Kegan Paul, 1986): 147.
42. Bruce Douglass, "The Common Good and the Public Interest," *Political Theory* 8 (February 1980): 105.
43. On the notion of a practice as a human activity whose goods are internal rather than external to it, see Alasdair MacIntyre, *After Virtue: A Study in Moral Theory* (Notre Dame: University of Notre Dame Press, 1981). On the idea of humans as "members and citizens" of a shared kingdom rather than as "conquerors" of it, see Aldo Leopold, *A Sand County Almanac* (New York: Oxford University Press, 1949), p. 204.
44. Environmentalists will recognize this language as that of Leopold (1949), pp. 224-225.
45. Galbraith, quoted by George Anthan in the *Des Moines Register*, October 11, 1987.

Chapter 2

Against Herbicide Resistance (1990)

There's corn in the bean field,
Persnickety wants it clean.
I got these blisters on my fingers,
I got these cocklebursin my dreams. . . .
I been walkin' the beans, in the burnin' sun,
And it looks like I ain't ever ever gonna get done.

- Greg Brown [1]

I pulled weeds out of half-mile rows of soybeans on grandma and grandpa's farm long before I heard of the controversy surrounding herbicide resistance and genetic engineering. Twenty years ago, Gordie, Richard, Greg, and I "walked beans," not knowing that our fists and scythes were not the only means available to Grandpa for killing weeds. We knew little then about uprooting thistles with tractors and discs or about spraying chemicals onto mustard. We knew only that a cool thermos of lemonade and some stern looks from Mom would motivate our troop into action because every good Iowan hated volunteer corn and sunflower shoots. The hatred stemmed as much from the fact that the weeds made a field "look messy from the highway" as from the fact that they cut down yields; Grandma had aesthetic sensibilities as highly developed as any character in Greg Brown's song.

The chore of weeding has fallen squarely into uncle Harold's hands for the last two decades. He understands that shelling out money in February to buy soybean seeds with a strong tolerance for ("resistance to") the presence of herbicides will cut his weeding costs in July. By applying chemical weed killers from his tractor, he saves the cost of hiring high schoolers to walk the beans.

In 1990, Iowa farmers have safer toxins and more discriminating spray equipment available than they did in 1970. But they still attack weeds the way farmers have done for thousands of years: They plant their crop, see what weeds come up, and then try to kill the weeds with chemicals, rotary hoes, and high school kids. For millennia, farmers have started with seeds, selected those they believe will be most productive, hardy, or drought-resistant, planted them, and then tried to devise means to eliminate their competitors. "Pre-emergent" herbicides, applied to the soil at planting, allow farmers to prevent weeds from coming up, but this is just a variation on the theme. The tune stays the same: You start with seeds that have traits you desire and then you find chemicals or cultural practices to get rid of everything else.

The time-honored technique is about to be reversed. Scientific developments have made it possible to start with chemicals rather than seeds. Having identified the genes that allow certain herbicide resistant weeds to survive in the presence of specific toxins, scientists have successfully transferred the genes from the weedy species in which they naturally occur to tomatoes, tobacco, and petunias. Soon, genetically engineered soybean and corn plants may be commercially available, crops designed to flourish in the presence of synthetic compounds. Glyphosate is a weed killer known by the trade name Roundup, assigned it by its parent company Monsanto. Glyphosate kills virtually every plant it touches. Starting with Roundup, scientists designed seeds to grow in the presence of the chemical.

Genetically engineered herbicide resistant (GEHR) crops reverse the order of weeding. Where our great great grandparents started with seeds and then hunted for chemicals, scientists now hunt for a chemical and then look for seeds. The reason is that a new generation of poisons has been discovered which seem to be far safer for humans and the environment than older poisons. If the new generation of broad-spectrum chemicals truly is safe, if you can drink glyphosate from a glass as proponents have done at press conferences, then seeds genetically engineered to grow in the presence of such benign chemicals would be welcome developments indeed.

That is the promise of GEHR research. Farmers, seeing ever tougher species of mutant weeds appearing in their beans, will have more efficacious and safer chemicals. Consumers, worried about pesticide residues on and in their vegetables, fruit, and meat, will have produce grown with less dangerous herbicides. Despite the excitement, however, there are problems associated with the technology. Leaving aside for the moment agronomic questions like whether GEHR crops will actually work in the field or how long it will be before weeds resistant to the new chemicals appear, consider the ethical questions.

Some critics have expressed reservations about the moral propriety of crossing unrelated plant species. Jeremy Rifkin, for example, has argued that it offends God to cross plants with weeds when the two species cannot be crossed by natural means of reproduction.[2] Is it right to violate species boundaries set up by "natural law"? This question may appear extreme to some plant geneticists and breeders, but it deserves the attention of moral philosophers interested in agriculture.

Others have expressed concerned that new labor-saving technologies may displace farmers. GEHR crops might increase the productivity and efficiency of an hour of a farmer's time, but what would that mean for farm and rural economies that are already unstable?

Some have worried about the medical and environmental safety of the final product. Will GEHR potatoes *really* be safe for humans, or will toxic residues remain in or on the vegetables? Will toxic compounds accumulate in the tissues of fish in streams collecting GEHR runoff?

Another worry concerns the economic power of the large chemical firms investing in GEHR crops, powerful multinational companies like Monsanto and du Pont. Will this technology allow a few chemical companies to strengthen their hold over an industry that is already oligopolistic, forcing American farmers to pay inflated prices for seeds and chemicals?

Finally, some are worried about who we are as a people, our communal identity. GEHR crops might make American agriculture more dependent on chemical-intensive and capital-intensive practices. Do we want to encourage exploitive attitudes to nature?

Ethical questions cannot be answered by scientific analysis alone. Ethics requires facts, and no one who closes their eyes to the science involved in agricultural biotechnology will be able to make justifiable moral decisions about it. But scientists can at best give us accurate descriptions of problems. Ethical judgments require philosophical reflection having to do with prescriptive analysis. Where scientists ask "What *is* going on?" and "What *can* be done?" philosophers ask "What *ought* to go on?" and "What *should* be done?"

My discussion has three parts. First, I introduce the difference between science and morality in general, and between natural laws and ethical laws in particular. This section is followed by a brief history of weeds and the techniques traditionally used to fight them.

In the third and longest part of the chapter I describe possible responses to the question of whether we ought to use public funds to do the research necessary to produce GEHR crops. There are four possible responses to this question. The first is *unqualified endorsement*, the view that GEHR research should proceed unimpeded if only because it is promising science. I analyze arguments that might be offered in support of this position and conclude that it is not defensible.

A second response is *unqualified opposition*, the view that GEHR research is not morally permissible under any circumstances. After analyzing arguments that might be offered in support of this view, I conclude that it is no more defensible than unchecked endorsement. In the last section, I offer assessments of two final views, *qualified endorsement* and *qualified opposition*. I conclude by arguing for the last position.

1. On science and ethics

Weeds may seem the stuff boring summers are made of, but reflection upon them uncovers a large and intriguing web of philosophical relationships. Consider the following epistemological principle: There is no such thing as a weed-in-itself. A virgin prairie does not in itself contain desirable and undesirable species; it merely presents diverse broadleaf plants and grasses competing and cooperating for nutrients. Not until an enterprising farmer or gardener or homeowner arrives are some plants suddenly transformed into "weeds" while others are turned into "turf," "crops," or "flowers." Weeds exist as much in the eye of persnickety beholders as they do "in nature."

Nature, for instance, does not call the dandelion undesirable (and neither does the farmer who grows it commercially for sale to wine makers). Those of us with a particular, historically-conditioned, landscape aesthetic select the dandelion for condemnation. According to another not implausible philosophical metaphor, then, Nature is a passive observer of the prairie, content to watch as the more fit outlast the less.

We are the ones to interfere, coming to the rescue of those plants that have garnered our favor. Rising above Nature, as it were, we insistently seek to bend her rules to our purposes. We intervene, saving the weak tomato seedling, pulling quackgrass, over and over again if need be, from a spot it may have occupied for centuries.

Seen in this light, weeding is more than a farm chore; it is an analogue of moral activity. For we are self-moving beings, possessed of a free will. We can move ourselves in directions contrary to our instincts. This difference between the active choosing self and the more passive dimensions of Nature identifies the difference between science, which studies the laws we call "natural laws," and ethics, which studies the laws we call "moral laws." Natural laws simply describe what happens. The law of gravity tells us about the relationship between two bodies, enabling us to explain what happens when a plant dies and to predict what will happen to its petals once they lack support. They will, of course, fall, inevitably, because of the laws Sir Isaac Newton discovered. However, as C. S. Lewis helpfully puts it, the dead petals are under no orders to fall, they feel no compunction to head downward.[3] They simply fall, time and time again, because of the attraction between physical bodies. The law of gravity describes this regularity, but it does not tell dead petals what to do next. Science describes.

Ethics prescribes. Where science explains and predicts, ethics compels and justifies, telling us not what is done but prescribing what ought to be done. When Mom asks me to walk beans, I can choose not to do what she has requested. If I refuse, I may feel that I have failed in my duties, that I *ought* to have obeyed. Unlike true natural laws, which cannot be broken, moral laws can be disobeyed. True moral laws, if there are any, are different from apparent moral laws, of which there are many. True moral laws over ride all other inclinations, binding us.[4]

When things are not the way they should be, we should intervene, trying to make things right, fair, just. In the 1980s, foxtail, quackgrass, cocklebur, velvetleaf, pigweed and other plant nuisances cost Iowa farmers approximately four percent of their potential corn harvest. In areas like the Mississippi Delta, where agricultural income is pinched at least as much as it is in the Cornbelt, losses from weeds were five times as great, averaging 20 percent of the possible yield. During times of great stress owing to unstable farm income, high input and capital costs, low commodity prices, and narrow profit margins--when the health of farm families as well as the diets of children around the world fare badly--the needs of humans clearly outweigh any alleged right to life plants may have. From an anthropocentric perspective, one that puts the needs of human beings above all others, weeds may be evil both in an agronomic and an ethical sense. To the extent that they are yield-reducing, they deprive children of food and farmers of profits. From this perspective, farmers are morally justified in trying to eradicate them.

But all of them? Some "weeds" may be useful because they prevent the growth of more harmful weeds, or because they contribute biomass to the field. As Levins and Lewontin explain,

> By using broad spectrum herbicides, beneficial weeds, those that compete with harmful weeds, are destroyed along with the harmful weeds they displace, so the weed problem is partly created by the very operation that is supposed to cope with it . . . The greater the cure, the greater the problem.[5]

Beneficial weeds are not thieves from an economic or horticultural perspective, stealing sunlight, water, and nutrients from desirable species. Ought they then to be regarded as ugly blots on a bean field? Concern for the "looks" of a field shows that considerations of *form* are sometimes as powerful in determining farm practices as considerations of *function*. Anyone interested in changing farm practices must therefore pay attention not simply to the science and economics of agriculture, but to its aesthetics as well.

The complicated web of weeds requires that a truly objective view of the subject take up issues relating to the anthropology of food, the sociology of scientific research, and the ideology of technological progress. Because food is such a constant part of our lives, cultures throughout history have expressed concern for its purity, going as far as to attach religious proscriptions against the eating of certain kinds of foods, or foods prepared or ingested a certain way.[6] To put those problems in proper context we need some historical perspective.

2. A brief history of weeds

Before the advent of modern herbicides farmers used organic techniques to control weeds. Thousands of years ago, in the first century of the common era, Pliny observed that the leaves of some plants could be used to discourage the growth of weeds.[7]

When hand hoes were supplemented with oxen and plows sometime after 1000 B.C.E., sound ecological principles were not abandoned. In Japan, traditional farmers grew rice in flooded river plains for centuries without depleting the soil. Sowing rice seed onto a bed of composted material, transplanting the seedlings into a field, and then lightly cultivating, hand weeding, and mulching, these farmers kept weed outbreaks to a minimum.

In the United States, farmers brought rotation schemes with them from Europe. Wheat, oats, and barley were often rotated over a five or six year period with corn and beans interspersed with years when the land would lay fallow or be used for pasture.[8]

Combined with mechanical cultivation, cultural practices were successful in keeping growers ahead of problem weeds. But insects were another problem; they cannot be controlled as well by cultural or mechanical means, and a single infestation of persistent worms or moths may devastate an entire field. Farmers waited until the late nineteenth century for chemicals that could deal effectively with insects and be produced on a large scale.

New machines, improved roads and railroads, and the adoption of "the revolutionized agriculture of England, with its scientific crop rotation schemes and conscientious application of fertilizers" induced farmers to specialize and intensify their cropping methods.[9] The best way to increase profits seemed to be to mass produce that one crop in which one had a competitive advantage. So tobacco farmers grew more and more tobacco and less and less wheat and beans while wheat and beans farmers concentrated on their specialties without much attempt to break into the tobacco market.

Intensive monocultural cropping practices led to profitable operations for many farmers, and they in turn came to depend on cash crops. Their dependency only deepened as the chemical industry discovered new compounds to battle new infestations of new grasshoppers, bugs, moths, worms, beetles. And weeds.

Arsenic was first used by an American farmer in the summer of 1867 to kill the Colorado potato beetle, which found ample and rich feeding across the Midwest and into the East. Farmers bought hundreds of reputed "solutions," but only one seemed to work both cheaply and easily: Paris green, a copper acetoarsenite. Within years the heavy metal was dusted not only on potatoes, but on "melons, squash, cabbage, a few other vegetables, most fruits and cotton."[10]

The results, as many writers then and now put it, were astonishing. The chemical industry had helped farmers take much of the risk out of farming and they in turn became remarkably productive. Using animal-powered implements in the early nineteenth century, American farmers could feed their own family of, say, six people. Using tractors in 1947, the same farmer could feed sixteen people.[11] The bugs, it seemed, had been beaten, and twentieth century farmers were hooked on toxins.

The chemical weed killers used in the late nineteenth century were inorganic compounds such as salts, oils, and chlorates.[12] When these caused visible harm to crop plants and, more infrequently, to farm animals and humans, the poisonings were taken as evidence that the compounds were efficacious. They were taken only secondarily as evidence that the product had an upper limit of safety to humans, and then farmers seemed to think only that the product had to be handled carefully. The high incidence of reactions to chemicals during the early period did not drive numbers of farmers to reject them; it seemed rather to reinforce the message sent by scientists and industry officials. Normal use of the chemicals posed no danger; it was only misuse that constituted a threat.

The chronic effects on humans of long term exposure to arsenical insecticides was of little concern, as was the accumulation of salts in the soil or the potentially toxic effects of compounds like PCBs (which were added to the oils to make them less flammable).[13] The desire for effective, easy to apply, chemicals motivated industrial research and development into cheaper, nonoil-based, compounds.[14]

Nonetheless, a grain of uncertainty about the agricultural establishment had been planted in the collective farm consciousness. This can be seen by considering another episode from the beginnings of the pesticide industry.

In the teens and twenties, farmers welcomed lead arsenate as a replacement for the old copper arsenite (Paris green) and calcium arsenite (London purple). The "new generation" chemical proved extremely effective against the gypsy moth, and was applied to celery in Florida in 1925 when growers there faced a plague of yellow-green caterpillars, on strawberries and pears and apples from Washington to Maine, and on grapes in New York.

Farmers would not learn how dangerous these chemicals were for many years. Arsenic and lead in small doses can cause symptoms such as "gastroenteritis, chiefly in the form of vomiting, diarrhea, and abdominal cramps" and, in only slightly larger exposures, neuritis, kidney and liver damage, "bloody urine, stools, and/or vomitus, cold sweats, thready pulse, and great prostration," even death.[15] Arsenic and lead are two of the most toxic compounds known, are virtually nondegradable, and are not soluble in water.[16]

Neither scientists nor farmers knew of these unintended effects when the arsenicals and metals were first introduced. But had researchers been more careful in their early work, they might have identified some of these effects. Fifty years before the introduction of the arsenicals, Townend Glover had reported retarded growth for plants grown in arsenic treated soil. Puzzled by his findings, he referred his 1870 study to the chemist William McMurtrie. Five years later McMurtrie reported little cause for alarm: plant growth was inhibited only after the amount of poison in the soil reached 500 milligrams, an extraordinarily large amount. Translated into field practices, this meant that it would take more than 906.4 pounds per acre of arsenic before soils would be damaged. Such a level of application, McMurtrie rightly pointed out, "would never be approached in practice."[17] The obvious conclusion was that farmers had no cause to worry about the safety of Paris green.

Even after farmers had stopped using London purple after seeing it damage their crops they stepped up their use of other arsenic based compounds. They were obviously prepared to avoid only those chemicals that did not work, not insecticides in general. And even farmers skeptical of the new compounds saw neighbors substituting the chemicals for expensive costs associated with farm labor, land, and management and were not blind to the economic advantages they offered. If resisting the "modern" way was not already nearly impossible, it was made even more difficult by the fact that few Ph.D.s in the research establishment had any use for the old ways. To argue against chemicals and for rotations in the 1930s was unheard of; who wanted their ideas labeled as attempts "to turn the clock back"?

There were, however, minority voices from time to time. As early as 1865 a journal called the *Practical Entomologist* dissented from the majority view, claiming that

> if the work of destroying insects is to be accomplished satisfactorily, we feel confident that it will have to be the result of no chemical preparations, but of simple means, directed by a knowledge of the history and habits of the

> depredators.[18]

Similarly, editors at the *British Medical Journal* wrote in 1892 that

> The use of poisons for the treatment of food is a matter which calls for the closest attention and the strictest control, where it is not absolutely prohibited under severe penalties. As in the case of food adulteration, the public cannot be left to the tender mercies of the interested or the ignorant.[19]

Such views did not carry the day, and the *British Medical Journal* was out of business within a few years. Instead, journals carrying accounts of miracle chemicals like those written about by McMurtrie proliferated. In one, A. J. Cook, a Michigan entomologist, reported in 1889 that he had sprayed one of his own fruit trees with double strength London purple until the liquid, Cook wrote, dripped onto the grass. In a remarkable move obviously intended to impress his audience, Cook then "cut the grass and fed it to his own horse. As presumably expected, the animal survived in good health, as did the sheep on whom the experiment was repeated."[20]

Once farmers had gotten used to dusting first arsenic and then DDT on their crops to protect them from insects, it was a short step to spraying them with herbicides to kill weeds.[21] This step occurred in the 1940s and 50s, and it was a revolutionary change. It is important not to overestimate the importance of synthetic agricultural chemicals before the mid-1940s. In 1876 there were only a half dozen herbicides in use; that number was almost unchanged in 1936.[22] Petroleum oils and salt products were in use before World War II, but mechanical means of cultivation were still the primary tool as late as 1939, when

> only about 30 [pesticides] were registered for use in the United States. Application techniques were limited largely to small sprayers and dusters, and applying pesticides was time consuming and the acreage of crops treated relatively small.[23]

But it is difficult to overestimate the importance of synthetic chemicals by 1950. More than one farm boy was surprised to return from the War to find his father's new sprayer behind the shed. How was such a rapid change possible? History again gives the answer.

While weeding has long meant stoop labor and drudgery, the expenditure of human energy on the control of weeds in chemical agriculture amounts to no more than 5 percent. Mechanical energy accounts for about 40 percent, while chemical herbicides take care of over half of all the energy needed to control weeds.[24] Herbicides replace not only human labor but the costs associated with mechanical tillage, fertilizer use ("without weed control, farmers would be fertilizing the crop and the weeds"), irrigation, crop yield losses, harvest costs (weeds clog equipment), grain drying (green weeds are high in moisture content and prolong the period required for drying the crop), transportation, storage, and land (herbicides reduce the number of acres needed for crop production).[25]

In 1945, the US already had at least fifty years of experience in dealing with and regulating agricultural chemicals. The first generation of chemicals were not widely used in the late nineteenth century. But the next generation (sodium arsenite and carbon disulfide, for example) attracted much wider attention from early twentieth century farmers. By 1914, petroleum oils were widely used to control weeds in irrigation ditches and in carrots.[26] So if industrially produced fertilizers and insecticides were not used on a large scale in 1939, cultural mechanisms and attitudes that would lead to eventual adoption were firmly in place. Traditional methods would prove to have little hold on farmers wanting to upgrade their operations to match the modern farms pictured in the magazines.

When synthetic herbicides came to market in the 1940s, American farmers were well acquainted with the use of chemicals. "Spray . . . spray . . . spray" wrote E. G. Packard in the *Entomological News*, and spray they did, since spraying relieved them of much of the worry and complexity previously associated with controlling pests. As Whorton puts it, "Once converted to arsenicals, farmers devoted themselves to the cultivation of ever better gardens with a Panglossian optimism that assumed that spraying could bring only good."[27] A cheap supply of sprays was assured because of the rapid development of the chemical industry during the war. Assisted with government loans, the industry had built massive production plants which were turned, after the war effort, to producing a cheap supply of pesticides and nitrogen fertilizers. Farmers found that spraying increased yields dramatically, thereby justifying the cost of the expenditure.

With the introduction of hybrid seed corn in the 1930s, pesticides induced farmers to plant more acres in monocultures. As mixed farms became rarer, grain farms became more specialized, and animals slowly came to be concentrated on feedlots. Consequently, less manure was available for fertilizing the grains. A twin set of problems arose. With fewer and fewer nutrients naturally present in the soil, more and more undesirable plants popped up. Farmers needed more nitrogen and more herbicides.

Answers to both problems came, once again, from the chemical industry. Fertilizer production was expanded. Between 1949 and 1968, food production increased by roughly 45 percent in the US while use of nitrogen fertilizer increased more than 600 percent.[28] Another "new generation" of herbicides was introduced: 2,4-D and 2,4,5-T and the rest of the phenoxyacetics hit the market in 1945.[29] These chemicals were used to control broad-leaved grasses on vacant lots, rangelands, and airports in addition to cropland. As the sheer quantity of chemicals produced grew, the prices to the farmer fell. Farmers responded by buying even more; a period of "acute food shortage" caused them to expand their productivity "as rapidly as possible at almost any price."[30] An expanding domestic demand for food and an ever growing export market to Europe encouraged farmers to produce as much as possible. They found that the "fastest, cheapest, way" was to increase their use of chemical inputs.[31]

By 1949 farmers were using 25 different herbicides on 23 million acres of corn, wheat, and turf. By 1959, one year after the introduction of atrazine, the number of chemicals had quadrupled, and the number of acres treated had doubled. Still, the 52 million acres receiving herbicides in 1959 represented less than 15 percent of total cropland in the US. The explosion occurred in the 1960s. Alachlor was introduced in 1969, and by 1974 over half of all crop acreage was receiving herbicides, a total of more than 160 million acres. The percentage of money spent on herbicides has also constantly increased. Whereas nitrogen and insecticide costs were dominant in 1951, 58 percent of a farmer's expenditures on chemicals went to herbicides in 1974.[32] By 1978, the tonnage volume of herbicides sold by the agrichemicals industry was second only to that of fertilizer.[33]

As herbicide use has gone up, so have total yields of crops and total values of crops lost to weeds. According to one estimate, 100 million bushels of soybeans were lost in 1970, a typical year, because of competition from weeds. This was the equivalent of what would have grown on 4 million acres.[34] As the value of crops lost to weeds went up, so did farm purchases of herbicides. By 1974, farmers were spending over one billion dollars each year on different chemicals designed to kill weeds.[35]

Why does herbicide use keep increasing? One reason is selective pressure. Herbicides wipe out a large percentage of targeted species of weeds, but they do not kill all of the individuals in any species. Some biotypes within the targeted species have a higher tolerance to the chemical. They survive the application, and reproduce quickly in fields where more fit competitors have been removed by the herbicide. Together with the fact that there are likely to be some weed species that are not killed by the herbicide, the fact of differential tolerance within species makes it necessary for the farmer to begin using more and different herbicides in succeeding years.[36] Each "new generation" of chemicals is soon met by species of chemical-resistant weeds, much as each new generation of insecticides is eventually confronted with mutant bugs that can tolerate the bug killer.

History shows that modern chemical agriculture is not a random system in which you can choose to adopt certain strategies and ignore others.

The rapid expansion in the use of herbicides after World War II went hand in hand with the use of industrially produced pesticides to control insects. Synthetic anhydrous ammonia--and now ureas--are used to supply nitrogen. Manufactured super-phosphates provide phosphate. And the system requires large amounts of capital to purchase the inputs, and large tracts of land over which to spread the costs.

Modern agriculture is an example of what Charles Perrow calls a complex, tightly linked, system.[37] As commercial nitrogen is used to stimulate the growth of high yielding varieties, it stimulates the growth of weeds as well. In 1965, corn farmers applied 75 pounds of nitrogen per acre. In 1987, they were using over 130 pounds per acre.[38] Herbicides are needed to control the weeds, and insecticides are needed to control pests introduced from abroad through internationally connected markets. The technologies used are increasingly expensive (a pound of atrazine sells for about $2.40, the newer alachlor for about $4.50, and glyphosate for approximately $22.00), so that farmers must have access to increasing amounts of capital for operating expenses.[39] And the farm input industries must make increasing expenditures to find new chemicals to deal with new pests.

You cannot play just one part of this game; if you use 2,4-D to control weeds, sooner or later you will need insecticides to control corn-leaf aphids stimulated by the herbicide.[40] Sooner or later, chances are that you will also need fungicides to control smut and Southern corn-leaf blight that also seem to accompany 2,4-D use.

Between 1951 and 1960, US farmers playing this increasingly close knit game lost an estimated $4 billion dollars per year to plant pathogens of one type or another. For all farmers, losses averaged 14 percent of the potential crop yield each season.[41] Farmers might have resorted to mechanical methods to deal with these losses but for the fact that expanded farm sizes mitigated against this choice as early as 1960. The invention of larger and more powerful machinery induced farmers to buy larger tracts of land so as to operate big tractors more efficiently. But once a farm had reached a certain size it became virtually impossible for the owner to substitute cultural means of weed control for herbicides. Given the sheer expanse of the fields, walking or cultivating the beans would prove to be prohibitively expensive when compared to chemical means of control. While we should not discount explanations that would focus on the natural tendency to choose a method that involves less physical labor, many farmers may have been using chemicals less by choice than necessity.

And they were not always using them wisely. In order to maximize yields and decrease risks, farmers tend to put on a little extra when a little less may be called for. Current levels of chemical use are not only "greater than the private optimum levels for plant nutrition and protection," but they are also in excess of what producers would choose to use if they were to maximize *profits* instead of yields.[42] The modern agricultural system depends on the pesticide industry and, in such a tightly linked system, farmers ignore chemical means of controlling pests at great peril.

In addition to showing why our modern farm system has become troublingly dependent on herbicides, the history of weeds suggests other lessons. First, the chemicals now in use in developed countries seem to have grown progressively safer for humans. Governmental regulations have tightened, and the arsenicals, the DDTs, and the oil based organochlorine pesticides have been banned for use in the United States. Second, as new chemicals come along, invariably accompanied by rhetoric about how "remarkable" the "new generation" is, our concern gradually shifts from the safety of consumers to the safety of farmers, farmworkers, their families, and workers in chemical manufacturing plants. That this is a smaller group than "consumers" in general is a welcome development, and we should not ignore the advances of the chemical industry in identifying newer and safer chemicals. But neither should our concern for the individuals affected decline. The individuals affected worldwide are not a small group since chemicals banned in the US are still made here and sold to developing countries. Third, as our concern shifts from ourselves to others, our environmental consciousness also expands, extending to ever larger portions of the animal and plant kingdoms.

The chemicals now in use in the developed countries are safer for mammals and humans than arsenic. As I just suggested, too many of us underestimate the moral achievement represented by the chemical industry's progress in the area of safety. But if new chemicals mean that many of us are increasingly safer and better fed, it also means that we are freed to consider our responsibilities toward less advantaged peoples, toward future generations, toward the environment, and toward the sort of people we want to become in the future. As we take up the new technology of genetically engineered herbicide resistant crops, these are the issues most needing attention.

3. Genetically engineered herbicide resistant crops

The idea of selecting for herbicide resistance is not new. Even before gene-splicing, researchers used traditional breeding techniques and mutogenesis to create new varieties of wheat with increased resistance to an s-triazine herbicide, terbutryn. In mutagenesis, wheat seeds are soaked in ethyl methanesulfonate and then grown in soil treated with the herbicide.[43]

The herbicide kills most of the seeds planted. But, on occasion, a very few mutant individuals survive; they become the basis for the herbicide-tolerant variety.

Genetic engineering speeds up this process. Researchers in Canada took weeds most resistant to the herbicide atrazine and genetically transferred their resistance to rapeseed and rutabaga plants.[44] DuPont has bred tobacco plants resistant to its sulfonylurea compounds, and Calgene has bred tobacco and tomato plants,[45] resistant to Monsanto's glyphosate, while Monsanto also has petunia and tobacco plants ready to go with its popular chemical.[46] Forestry and chemical lawn industries are watching with great interest as private labs and public universities apply more and more sophisticated genetic engineering techniques in herbicide resistance research. Much of the research is funded with public tax dollars, and much of it is going on at land grant universities whose charge is, in part, to educate and help to improve the well being of "the industrial classes."

Ought we to use taxpayers' monies to fund this research? To answer such ethical questions, we must have ethical principles to guide us.

4. An environmental perspective: Three principles

Moral questions call for public debate in which neither scientific expertise nor philosophical acumen necessarily constitutes authority. Because these are communal issues regarding how we ought to relate to each other politically and socially, authority in the debate does not come from technical competence. It comes from practical wisdom, what Aristotle called *phronesis*, the ability to see which response fits the occasion. But which response to publicly funded GEHR research "fits the occasion?"

When we encounter disagreements over morality, it helps to begin with judgments about which we have a firm consensus. We agree, for example, that it would be wrong to genetically engineer children to live lives of unrelieved, purposeless suffering. We have a firm consensus that it would be wrong for a company consciously to genetically engineer seeds that would produce crops with high levels of undetectable carcinogens.

In those areas where we have firm moral concensus, we erect laws and make it illegal to do what we think plainly immoral. It is important that we agree on fundamental values such as the worth of human life and the importance of freedom of expression and inquiry. If we truly lacked a consensus on such basic matters, conversation and communal life would be impossible. So, even when different moral traditions clash over particular issues, there is almost always a residual overlap of what John Rawls calls considered convictions.[47]

GEHR crops bring up moral issues in three areas: our duties toward the natural environment; our political and economic responsibilities; and our ideals and attitudes as a community. I do not propose the following principles as moral truisms, but rather as principles that seem, from my perspective, the most defensible in each area.

Regarding the environment, I believe that Aldo Leopold's basic principle of the Land Ethic is apt, that "a thing is right when it tends to preserve the integrity, stability, and beauty of the biotic community [and] wrong when it tends otherwise."[48] The way we might put this principle is as follows:

> 1. A principle concerning the environment: *We ought to show respect for different values of different things.*

This principle enjoins us to treat beings in ways appropriate to each species' ecological niche and each human's individual rights. Rational agents and patients are entitled to be treated as ends-in-themselves. This means that they have basic rights to life, food, shelter, and autonomy. Respect here means doing no harm to individuals, making efforts to secure their well-being, and maximizing possible benefits and minimizing possible injuries to them. Sentient animals are not autonomous agents, but their ability to experience emotions, memories, and desires, and to form intentions, families, and societies entitles them to be spared unjustifiable pain and to be left alone to flourish or die according to laws operating in their ecosystems. Nonsentient animals, plants, and organisms have a lower level of conscious awareness, if any, but their importance to the ecosystem entitles them to be treated in ways that will preserve nature's beauty, integrity, and diversity.

My environmental principle endorses a hierarchical view of moral value, and seeks to establish value distinctions on the basis of the complexity and intensity of experience. Mine is not a strongly anthropocentric theory in which the value of any particular thing is tied to its value for humans. On my view, plants, organisms, and animals have value in themselves. This environmentalist view is not anti-anthropocentric. It is an environmental ethic in which the basic rights of individual humans (see below) takes precedence over the good of individual animals.[49]

Regarding the political economy, I believe that efficiency and economic growth are defensible values. Our particular history of prejudice and discrimination in the United States, however, requires that we put equity and equality of opportunity at the top of the agenda, above efficiency or growth. Thus:

> 2. A principle concerning the political economy: *We ought to pursue economic justice and equity first, and economic efficiency, productivity, growth, and entrepreunership second.*

This principle enjoins us to put first the need for economic opportunity by society's worst-off. An economic system coherent with this principle places priority on redressing persistent inequities resulting from discrimination on the basis of class, race, and sex. It also insists that external environmental costs, which may harm future generations of society's worst-off, be included in the cost of production. The principle enjoins us to maximize efficiency and productivity in the workplace while encouraging imaginative and fair competition among individuals and corporations only after attending to concerns relating to equitable opportunity.

Regarding our communal identity, I believe that we must take into consideration the wisdom of past generations and the needs of future generations. Thus:

> 3. A principle concerning our community: *We ought to form and maintain diverse, just, and beneficent, communities.*

This principle enjoins us not to pursue public policies likely to have the effect of standardizing society or benefiting certain politically powerful communities at the expense of marginal or sectarian groups. Rather, we should try to form diverse international communities in which the basic rights of all individuals and groups are met. This means paying attention not only to the development of social policy, but to the development of spiritual character. We ought to help our children, for example, to develop dispositions of proper humility before God, appropriate respect for Nature, appreciation for the wisdom of ancestors, and consideration of the needs of future generations.[50]

Not everyone will agree with these principles, and I state them not because I think they are noncontroversial but only so that my starting point is clear. Readers who already find themselves in agreement with my principles may want to see whether the conclusions I reach about GEHR research cohere with our shared principles. Those who disagree with my principles may wish to articulate their own moral foundations and then see what policy implications follow from them.

5. Four ethical responses to GEHR crops

5.1. Unqualified endorsement (E)

What should we say about the morality of using public monies to fund research on GEHR crops? The first position I want to consider is that GEHR research should be endorsed without reservation. Let us explore such a position.

Some think that all scientific research is morally justified insofar as it may lead to new knowledge. Indeed, the *potential* benefits of any scientific research project are huge. Potential spin-off products and discoveries from basic research are almost always unpredictable and are sometimes almost unfathomable. Think of the unforeseen benefits that came from research on penicillin. For this reason, it would be difficult to make the case that *any* scientific research program had a negative benefit/risk potential. In fact, few if any governmentally-sponsored projects have ever been thwarted on these grounds.[51]

The problem, of course, is that we do not know what knowledge may result from research. If we knew ahead of time which projects would make important advances and which would not, we could easily decide which ones to fund. Unfortunately, this kind of prediction is precisely what we cannot do. Consequently, we fund a variety of projects, not all of which give us as many desirable results as others. But this leads those in *E* to give unqualified endorsement to GEHR research simply because it seems as likely, if not more likely, to result in beneficial agricultural products.

Insofar as research is being funded privately, there is little one can say in rebuttal. Privately held corporations ought to be good stewards of their shareholders' monies, and so they should pursue those lines of research that they think may prove profitable provided, of course, that in doing so they do not break relevant moral principles. Because I value economic efficiency and productivity, I believe that if GEHR research has no other morally objectionable features, then the simple fact that private corporations may make a profit by researching it is not a reason for opposing it. It is a reason for endorsing it.

The question, of course, is whether there are *other* moral grounds on which to object to GEHR research. In order to pursue that question, I propose to bracket the issue of *privately funded* GEHR research, and to confine my attention to research being conducted with *public* funds.[52] This means research at land-grant and other public universities, at governmental agencies like the USDA, and at private corporations receiving federal or state research monies.

Those in *E* believe that GEHR research is defensible in part because scientists think it may advance scientific knowledge. But there is another defense: The world's population is growing dramatically while environmental conditions in many traditionally high-producing agricultural areas are becoming less hospitable to agriculture. If we are to meet the growing demand for food we will need new crops and new chemicals suited to growing conditions in new geographical locations, and soil and climactic conditions.

This defense of GEHR research contains two claims about the potential economic and environmental benefits of that research.[53] Those arguments must be individually assessed.

A. GEHR research will lead to economic benefits for manufacturers, consumers, and farmers.

There are obvious pecuniary benefits to be made by the agricultural biotechnology firms that successfully market GEHR seed and chemical packages. Monsanto, American Cyanamid, Calgene, FMC Corporation, and E. I. du Pont are the five major players, and each anticipates healthy returns on their research and development expenditures. The values of economic efficiency and productivity are genuine values, but do they justify GEHR research? The only way to answer the question is through ex-ante economic impact analysis, trying to determine whom the technology is likely to benefit and whom it is likely to hurt.

Multinational corporations stand to be the biggest winners, of course, but risks are attached. Research and development of any new chemical is very expensive, and the major players could stand to lose money if they encounter problems such as corporate mismanagement, product failure, stringent environmental regulations regarding field-testing or marketing of GEHR seeds, or inability to compete with other manufacturers.

Will a handful of companies establish monopoly-like control? By the estimates of Charles Benbrook and Phyllis Moses, there are more than two dozen "agrochemical and biotechnology companies" in the race to produce GEHR seed and chemical packages. This competitive atmosphere makes it likely that "the companies' prospects for [overall] sales increases [resulting from GEHR technology] are modest compared with the total volume of business conducted by them."[54] Moreover, competition from old as well as new compounds combined with farmers' natural tendencies to switch methods slowly makes it "not likely that the structure of agribusiness will change dramatically simply because of herbicide resistant plants."[55] There are grounds for thinking, then, that GEHR research will benefit manufacturers without substantially changing the current structure of the industry.

Will consumers join multinational corporations as beneficiaries? Slightly higher yields and slightly lower input costs for the farmer should lead to larger corn surpluses, and this will mean corn prices as much as 30 cents per bushel lower than they would be otherwise. Lower corn prices should translate into lower beef prices at the supermarket, and that would be good for domestic, and probably foreign, consumers.

In the short run, some farmers may also benefit financially from GEHR crops. If early adopters cut weed losses and improved the efficiency of pest control by substituting GEHR seed and chemical packages for existing weed control measures, they might capture competitive advantages lost by late adopters. GEHR crops should help these operators to reduce outlays not only for herbicides but for fungicides and insecticides as well, since weed-free fields might cut insect and disease problems.[56] It would also help to cut harvest expenses related to weed clogged machines and postharvest drying costs inflated by crops infested with moist green weeds.[57] In the short run, a small group of farmers may join multinationals and consumers as beneficiaries of GEHR research.

In the long run, however, all corn and livestock farmers who use chemicals seem likely to be financial losers as corn and beef prices decline. Farmers in different regions of the country and with different modes of operation will suffer to different degrees. On the one hand, the growing number of organic and natural farmers who eschew the use of synthesized chemicals will not be directly affected by GEHR crops because they will not purchase the herbicides. (But even they will be affected indirectly by falling farm gate prices.) On the other hand, corn growers in the Delta who use herbicides already suffer most from weed problems, and they may benefit in the long run as GEHR research products enable them to grow more corn. Finally, corn growers in the corn belt whose incomes derive substantially from this one commodity will probably suffer most from lower prices induced by GEHR technology.[58]

As the world marketplace shrinks, our farmers face increasing competition from foreign producers. Agricultural biotech products like GEHR crops could help our industry remain at the forefront:

> As the future wave of agricultural innovation, biotechnology promises to decrease the need for expensive agricultural inputs, increase production efficiency . . . and create new crops and livestock species as well as new products from

> current surplus commodities. . . . The United States can reinforce its world commercial position.[59]

For anyone concerned with the survival of family farms, these benefits should not be weighed lightly.[60] For anyone holding my political and economic views, the promise of an increased efficiency in the use of agricultural resources is a genuine advance.

On the basis of potential economic benefits and risks to those in developed countries, the only apparent reason not to endorse publicly funded GEHR research is the long range effect on farmers. But even that economic adjustment is one that I would be willing to accept as fair because it seems to be consistent with adjustments required of all sectors in competitive economic systems.

B. GEHR research will lead to seed-and-chemical packages that are safe for humans, animals, and the environment.

This claim needs to be carefully considered because increasing use of herbicides has already led to the detection of contaminants in surface and groundwater in rural areas. Alachlor and atrazine, for example, make up 36 percent of all the chemical weed killers presently used in the United States. They are known to cause cancer in animals when administered in high doses. And they are being detected in wells and surface water in rural areas of Iowa and other parts of the United States.[61]

The possibility is real that alachlor and atrazine are adversely affecting rural residents in agricultural areas. A study conducted in 1989, for example, concluded that there is a statistically significant increase in the number of underweight babies born to mothers in rural Iowa where drinking water comes from surface sources containing herbicides such as atrazine.[62]

GEHR crops might help to alleviate these problems insofar as many of the crops being designed are intended to be grown with the new, safer, herbicides. In addition to Monsanto's glyphosate, these herbicides include du Pont's sulfonylureas (trade name "Classic") and American Cyanamid's imidazolinones ("Scepter").

The sulfonylureas, imidazolinones, and triazoloprimidines kill weeds by inhibiting the formation of the acetolactate synthase enzyme (ALSase), a biochemical pathway in plants that does not exist in insects and mammals. Consequently, the chemicals appear to have few adverse side-effects on animals, and many scientists believe that they are far safer for humans, mammals, insects, and birds than the old halogenated herbicides, chemicals which have chlorine, fluorine, or bromine added to them in the manufacturating process. The new herbicides are composed solely of carbon, hydrogen, oxygen, and sometimes nitrogen, simple compounds that appear abundantly in nature.

In addition to being safer for humans, the new herbicides may be safer for the environment because, as Benbrook and Moses write, "the herbicides either break down rapidly into carbon, nitrogen, and oxygen in the environment, or they do not leach appreciably into water."[63] Further, "no unique environmental or ecological concerns have been associated with resistant cultivated plant lines, regardless of fears about genetic engineering."[64] Benbrook and Moses claim that the new chemicals do not move into water because the herbicides are rapidly degraded by plants, air, sunlight and soil. If all of these claims prove to be correct, the new generation might be far safer for farmers who must handle chemicals, and health problems related to herbicide use might decline.

If GEHR crops help us to lessen our reliance on old herbicides, this might be a distinct advantage. But we often do not know the dangers that come with longterm use of herbicides, as the history of 2,4,5-T use illustrates.

2,4,5-T was introduced in the 1940s. Unfortunately, it was not known then that the process of manufacturing 2,4,5-T can produce a contaminant called TCDD, one of the most toxic chemicals known. According to one estimate, TCDD is "150,000 times more toxic than organic arsenic," in addition to being teratogenic, the cause of birth defects.[65]

The tragic side-effects of 2,4,5-T may not have been known in 1958 when the US Defense Department purchased 5.8 million pounds from chemical companies. And they may not have been known ten years later, by which point the Department had increased its purchases more than 500 percent.[66] But we now know what TCDD is and does. A decade after the war in Vietnam, dioxin was still found contaminating the ground. The main hospital in the Tay Ninh region, northwest of Saigon, reported after the war that one out of four pregnant women seen in the hospital miscarried. During the period of 1968 to 1970, when huge amounts of 2,4,5-T were being used by the US to defoliate the forest in Tay Ninh, stillbirths were twice the rate of nonsprayed areas of Vietnam. There were other problems. In Saigon's Children's Hospital, the incidence of spina bifida and puracleft palate was three times higher during spraying than at other times, an estimate confirmed independently by the US Defense Department.[67] Wives of US soldiers also reported birth deformities in infants.[68]

2,4,5-T has been banned or restricted for use in the US since the mid-1970s. But it was still being manufactured and exported to developing countries as recently as 1978.[69] So concerns about its effects on those who handle and manufacture it, in addition to concerns about potential military uses, are not irrelevant for those interested in the moral dimensions of herbicide research, manufacturing, and use.

In the face of the toxins present in some of the older herbicides and nitrates deposited in water from synthesized fertilizers, the new generation of environmentally safer herbicides is a welcome development. The herbicidal action of the new generation apparently work by blocking the production of specific enzymes that make amino acids essential for the regulation of a weed's growth.[70] While some of the older herbicides also work on biological pathways not found in rodents and higher animals, the new herbicides continue this research trajectory.

The new compounds, for example, are only half as toxic to mammals as alachlor; whereas it would take only 1600 mg. of alachlor per kg. of body weight to kill 50 percent of a group of rats, it would take almost three times that amount of Roundup to kill the same percentage.[71] Not only safer for farmworkers to handle, they are also much more potent for plants, and in significantly smaller dosages. Glyphosate goes on at one fifth the rate (0.5 pounds per acre) of alachlor (Lasso requires 2.5 pounds per acre to get the same control), Scepter at one fortieth the rate (0.125 pounds per acre), and Classic at one eightieth the rate (0.03 pounds per acre).

Moreover, the new compounds are applied at greatly reduced rates. While alachlor ("Lasso") must be applied at 2.5 pounds per acre to be effective, and atrazine ("AAtrex") at 1.8 pounds, glyphosate goes on at 0.5, imidazolinone (in the form of imazaquin) at 0.125, and sulfonylurea (in the form of chlorimuron ethyl) at 003.[72]

These are the arguments of those who would give unqualified endorsement to GEHR research. But there are problems with the arguments. First, the new herbicides are not the only ones that work on biological pathways absent in humans and animals. Atrazine works by inhibiting photosynthesis, so *the way* in which it actually kills plants is not harmful to humans. Nonetheless, atrazine presents some potentially severe problems for us.

Second, it is a very difficult task to ascertain just how synthetic chemicals behave in farm fields. There are very few field studies of how herbicides, old or new, move through the rural environment. In part, this is because it is such a complex matter to identify the by-products of degradation. While carbon and nitrogen dioxide may constitute the majority of end-products, there are probably lots of intermediate products along the way. So to imply that the new herbicides immediately break down into three simple compounds, is to imply that we know more than we do know.[73]

Defenders of GEHR crops are not unaware of criticisms. To the worry that gene-splicing represents a new and strange historical epoch, they might reply that herbicide resistant crops are nothing new. Varieties have long been selected for their resistance to herbicide and insecticides; even before genetic engineering came along we have been identifying and marketing seeds that could grow in the presence of chemicals used to kill competitors. Genetic engineering only speeds up the process by cutting down on the length of time needed to come up with new varieties; what moral questions does it raise that could not be raised about traditional plant breeding techniques?

The environmental safety of deliberately releasing genetically engineered organisms and crop plants into the environment is also debated. Few, however, have argued as forcefully as Winston Brill that we have little to worry about in this regard. Confining his attention to genetically engineered organisms without saying much about plants, Brill argues that the changes being made are virtually negligible from either a genetic or an environmental perspective.[74] The reason is that

> the best a genetic engineer can do is add one gene, or at most a few genes, to the tens of thousands of genes in an organism's chromosome . . . A useful organism, therefore will not inadvertently be converted into a pest, pathogen, or entirely new species.[75]

Brill's point is that new organisms undergo such minor changes in their genome that they are not sufficiently different from naturally occurring organisms to constitute a threat. Indeed, Brill argues, these organisms are so fragile that they can barely exist outside of the laboratory, and in over fifteen years of lab experiments, not one example of a mutant organism-out-of-control is known.[76]

GEHR tobacco and tomatoes, or even soybeans and corn, would not be hardy enough to predominate in the wild. So Brill argues with regard to organisms. "I think it's going to be extremely difficult to [use genetic engineering to] make an organism worse than any organism we now have," he adds. Then he extends his argument about organisms to crops, concluding that

> absolutely no scientific basis exists to believe that by genetically engineering corn, wheat, or rice one could inadvertently produce a serious problem weed. Serious problem weeds are not the result of a change in a single gene. They must in general terms meet a variety of criteria. The seed, for instance, would need to survive for a long time; it might have to be dispersed over a great distance; the plant would have to grow faster and be more vigorous than the plants around it. These properties are not produced by one

> gene, but by hundreds if not thousands of genes. . . . How could one imagine that by engineering one or even several genes in an organism, corn might be converted into a problem weed? The chance of producing a problem weed through this technique is less than the chance of producing one through a traditional cross of corn and teosinte. And no one is concerned about the latter.[77]

If Brill is right, the deliberate release of genetically engineered crop plants need not concern us. The germplasm of the new varieties will be virtually indistinguishable from the crops and weeds currently found in farm fields.

Brill makes much of the scientific community's clean history of environmental releases, in which the past decade of experiments have placed billions of microorganisms into the environment without producing a single pathogen. Why concern ourselves with such a safe practice when traditional breeding methods have produced novel species for thousands of years without arousing moral indignation?

> The case of corn crossed [through traditional means] with teosinte . . . [resulting in the mixing of] tens of thousands of genes from each plant . . . [has not led anyone to take] special precautions because of the vast and safe experience with such crosses.[78]

While traditional breeding methods are safe, genetic engineering is safer because it is a rational procedure, "a far more precise and much more predictable process." Only one or two genes are exchanged, not tens of thousands. Moreover, such exchanges occur quite naturally, "between unrelated bacteria . . . between bacteria and plants . . . between animals and plants . . . and between animals and bacteria." And yet only the most fit of these mutants survive; "new dominant species" do not arise "routinely." So even though "millions of acres are treated with chemical herbicides . . . no health or environmental problems have occurred from the large numbers of . . . uncharacterized microorganisms, with unnatural and uncharacterized genetic alterations [induced by widespread herbicide use on] farms, golf courses, and gardens." Finally:

> even though these microorganisms can persist, transfer to distant sites, and exchange their new genes with other microorganisms, no agency has demanded tests for health or environmental problems from such organisms.[79]

Brill argues that the huge number of uncontrolled genetic mutations in nature dwarfs the controlled mutations engineered in the laboratory. History shows that the vast majority of natural mutations do not survive, and that those that do rarely succeed in disrupting the environment. The environment simply adapts to them and goes on.

One must agree that the history of deliberate releases, so far, has not yet produced virulent pathogens. But we have only been releasing genetically engineered organisms for a decade or so, and with extremely tight (some would say excessive) government regulation. Should we be so confident that in the future, when government supervision relaxes and the number of releases takes a sharp turn upward, that the safety record will remain unblemished? The problem is that deliberate release is a low probability but high risk enterprise; one mistake could have potentially disastrous consequences.

There are other problems with Brill's argument. How consistent is it to argue, on the one hand, that genetically engineered organisms are not hardy enough to predominate in the wild while arguing, on the other hand, that modified organisms are barely any different than their natural cousins? If the natural forms are hardy, and if the mutants have had only one gene replaced, then the fragility of the mutant shows how dramatic a single gene change can be. Why should we think that single gene replacements will make organisms less competitive? The point of deliberately releasing new varieties of plants, at least, is to demonstrate the opposite point: that the new variety is hardy enough to be grown by farmers "in the wild," as it were. If we knew ahead of time that the variety was so fragile that it would not be able to grow in the field, why would anyone want to test it?

Applied to GEHR crops, an argument about plants analogous to Brill's argument about organisms would have us believe that genetic engineering can produce good hardy crops but not good hardy weeds. Benbrook and Moses make this argument, claiming that resistance will not spread from crops to weeds and, if it does, farmers will just have to change management practices.[80] But, since we already know that a crop in one field is a weed in another, the argument that herbicide-resistant crops cannot result in a new generation of herbicide-resistant weeds is self-contradictory.

There is another way to read Brill's argument, and that is as an argument from evolutionary genetics and mathematical probabilities. In its millions of years of evolution, one might argue, nature has produced so many genetic mutations that all mutations adaptive enough to persist are already present in the environment and any novel genetic modification must necessarily be ill-adapted. If this is what Brill has in mind, then there are a different set of problems with the argument because, according to Philip Regal, mathematical genetics has disproven the idea that every genetic possibility with adaptive traits has already occurred.[81] Nature is not a closed system, and the past million years of evolution have not necessarily produced every viable organism, plant, or animal possible. So interpreting Brill to claim that no new hardy genetic mutations are possible is not an interpretation that will save his position.

Phyllis Moses and Charles Hess have asserted, like Brill, that the products of rDNA research are "no more likely to have suddenly acquired unknown and dangerous attributes than if created by breeding or cell culture, which, similarly, yield genetically altered organisms."[82] But given the sensitive and congenial laboratory conditions which welcome rDNA organisms into the world, one wonders how Moses and Hess can be so certain about this. After all, we have only been experimenting with these altered organisms for a decade or so. Indeed, after boldly asserting their claim in one sentence, the very next sentence implies that theirs is a hypothesis, not a proven fact: "Adequate testing of products derived by recombinant DNA methods should proceed swiftly to gain sound scientific evidence for this proposition." Does the second claim indicate that their prior assertion is not, contrary to the way it is couched, a statement of fact? Or are they simply impatient with the speed of scientific testing, which has not yet proved what they know to be the case? In calling for testing to "proceed swiftly to gain sound scientific evidence" are they announcing the silliness of such research, since they already know that the evidence is there? Or are these public officials trying to tell scientists what they ought to find? The precise meaning of their argument for the safety of genetically engineered field releases is as difficult to discern as Brill's.

In addition to the fact that it rests on an arguable claim for the utter safety of genetic engineering, another problem with *E* is its assertion that the new biology will usher in a new era of agriculture in which only the newest, lowest dose, safest, herbicides will be used. At this point I want to draw attention to the rhetorical techniques some scientists have used to argue their views. Science, of course, does not occur in a vacuum, and much depends on the ability of the scientific community to mobilize political support and financial backing for their projects. So an analysis of the ethical dimensions of scientific research is naturally drawn to the scientists' concern not only to inform but to persuade.

In this light, consider Benbrook and Moses' argumentative technique of listing propositions that appear to be true but, according to them, are in fact "myths:"

> *Myth 2* [their emphasis]: Progress in developing herbicide resistant cultivars will increase the use of chemical pesticides.

This myth is false, the authors claim, because "the whole thrust of resistance R&D propels US agriculture toward products effective at one-half to one-tenth or less the current rate of application of older products on a per acre-treated basis."[83]

We recall, however, that the authors have already informed us that roughly 75 percent of all GEHR research is directed toward finding crop varieties tolerant to the new chemicals. But that, of course, means that roughly twenty five percent of all GEHR research is directed toward finding crop varieties tolerant of the old chemicals.[84] Is it accurate, then, that "the whole thrust of resistance R&D propels U. S. agriculture toward" the new chemicals? Work is being done, for example, to manufacture soybeans with resistance to atrazine, which would allow farmers growing corn and beans in rotation to use stronger doses of atrazine on their corn without having to fear carry-over into next year's bean crop. One estimate has it that such a strain of soybeans would increase sales of atrazine by $200 million a year.[85] Does this justify the claim that GEHR research is likely to introduce a new era of agriculture in which only "safe" chemicals are used?

Another hidden economic cost is the likely appearance of herbicide-resistant weeds. Consider the number of dormant weed seeds already in the soil that already possess the ability to grow in the presence of some of the new herbicides.

> Inhibitors of the ALSase enzyme are predicted to occur in natural weed populations at a rate of . . . 1 in a million. Considering the vast numbers of dormant weed seed in agricultural soils, this is a relatively high rate of occurrence. Triazine resistant weeds occur much less frequently, about 1 x 10^{-18}. 2,4-D resistant weeds are probably even less likely to occur possibly due to the complexity of its action. The recent introduction of many ALSase inhibiting herbicides from several different chemical families could lead to a herbicide rotation with only one target site in plants . .[86]

The high rate of dormant weeds with potential resistance to the new chemicals leads Dekker to conclude that herbicide resistant weeds are "inevitable."

Here, then, are several hidden economic costs that need to be explored before concluding that the potential economic benefits of GEHR crops outweigh their costs. To assert that worries about the hidden costs of GEHR crops are "simplistic" is itself simplistic.

But suppose we grant the claim that GEHR crops will reduce external environmental and health costs. This still does not justify the assertion that genetically engineered plants are "safe." Like the meaning of the concept "weed," the meaning of "safe" is context dependent. Its meaning depends on a wide web of relationships which need to be specified each time it is used.

Safe for whom, and under what circumstances? Safe with what probability of error, and with what safeguards? Safe for how long, and with what risks of subclinical health problems? Is what is safe for a forty-year old woman in good health also safe for a sixty-year old man with an immune deficiency?

We have seen more than one instance of the scientific community declaring a new synthetically manufactured chemical to be "safe" only to learn later of unanticipated deleterious side-effects. One lesson of the history of weeds was that scientists should exercise great restraint in claims about the safety of new technologies. We must be more modest than Brill, Benbrook and Moses, claiming not that genetically engineered organisms or GEHR crops are safe but rather that they appear to be unlikely to disrupt the rural landscape much more than the crops and chemicals currently in use. We might conclude that some of the chemicals likely to be used with GEHR crops appear to be safer than the old chemicals insofar as their carcinogenic, mutagenic, and teratogenic effects on fish and mammals are concerned. And we might say that GEHR crops are, given what we now know, "safe enough" for this or that purpose, given these and those ends of modern agriculture. What we should not say is that GEHR crops are safe.[87]

Might the argument for unqualified support of GEHR crops be put in more defensible, contingent, terms? Might we say that GEHR crops *should* make it *possible* to promote a style of farming that uses *safer* chemicals that are *less* toxic and *more* compatible with the environment? We might, but we would no longer have the unqualified assertion of Brill's title, nor an endorsement of GEHR crops without reservation. For once appropriate relativizing clauses are inserted, *E* is no longer *E*.

Unqualified endorsement of GEHR research and crops collapses from internal problems. Key questions are begged, like those relating to the meaning of the concept "safe;" contentious claims are asserted, like those relating to the ability of the new herbicides to replace the old; and hasty judgments are paraded as arguments, as when legitimate concerns about hidden costs are dismissed as "simplistic."

There is little justification to endorse GEHR research without reservation. Ought we then to oppose it?

5.2. Unqualified opposition (*O*)

Judged by economic standards like efficiency and profitability of industries, farmers, and consumers in developed countries, publicly funded GEHR research has much going for it. But there are problems associated with the fit between GEHR technologies, the environment, and our communal identity, problems that are not obviously outweighed by the potential economic gains. Some critics have responded to GEHR research by uncategorically opposing it. Those like Jeremy Rifkin who have biases against genetic engineering, think that any kind of manipulation of life at the genetic level is impermissible. Others, including Jack Doyle, are concerned about the effects of a concentrated agribusiness industry on concerns related to equity and distribution of wealth. Others see danger in continuing to allow the chemical industry to locate its manufacturing plants in places such as Bhopal, India, where 2,000 people were killed in an explosion in December, 1984. Indeed, one estimate puts the number of pesticide poisonings at "between 400,000 and 2 million . . . worldwide each year, most of them among farmers in developing countries."[88] Others estimate that there are between ten and forty thousand deaths resulting from these poisonings each year. In light of the ethical directive to do no harm, should we continue to pursue technologies that will perpetuate pesticide production and use?

Those who think that we should stop research on GEHR crops do so for at least four different reasons. First, they argue, the research will lead to an increased use of chemical pesticides, and more farmworkers and consumers will be injured or killed as herbicide use escalates. Second, mutant organisms may develop in GEHR crop fields and devastate vast areas of vital crops. This would put the food supply of the entire world at risk. Third, a small handful of companies may exploit farmers and consumers by exercising monopolistic control over the seed and chemical industries. Fourth, some are convinced that GEHR research is intrinsically immoral because it crosses species boundaries placed in nature by God.

The first two arguments that might be offered by those in *O* can easily be shown to suffer from the same problems encountered by arguments offered by those in *E*. Consider that there is an active debate about whether GEHR crops will actually lead to an increase in the use of the old herbicides.[89] According to Benbrook and Moses' estimate, three quarters of current expenditures on GEHR research involve the use of low-dose herbicides like Scepter and Classic, chemicals applied at rates that are a fraction, one fifth, one tenth, one hundredth, of current rates.[90] If these become the chemicals of the future, then the rates of herbicide use may indeed fall. In fact, that would only continue the pattern that has prevailed recently. Application of insecticides in general, and of herbicides for dicot control in cereals in particular, have declined steadily since 1940.[91] At best it is unclear whether herbicide use will escalate, and it is possible that use will level off or decline with the advent of GEHR crops. To oppose GEHR research on the basis that GEHR crops will increase herbicide use is to argue on shaky grounds.

The safety of the technology for farmworkers, manufacturing employees, consumers, wildlife, and the environment is also debated. Just as those in *E* ought not to assert that GEHR crops will be "safe," so those in *O* ought not to assert that they will be "unsafe." The problem here is not that we cannot predict the future, a problem that plagues anyone trying to anticipate the likely effects of a new technology. The problem is rather that the new generation of herbicides has proven to be far less toxic not only than the old organochlorine insecticides like DDT and paraquat, but less toxic than the most popular herbicides now in use, alachlor and atrazine. It is as difficult to defend the claim that GEHR chemicals and crops are unsafe, or that they will lead to more health, environmental, and manufacturing accidents, as to claim that they are safe and will solve all of our problems. Our judgments must be more qualified and specific, taking comparative forms like "safe enough for muskrats but not for human babies," "riskier than what we now have for worms, but much less worrisome for field workers."

Regarding the potential impact of GEHR crops on the appearance of the rural landscape we must say that it is possible, although probably not likely, that GEHR plants may cross-breed with wild weed species. But even if this occurs, grand ecological disaster will not necessarily follow. As Benbrook and Moses suggest, farmers will adapt once again, inventing new strategies for dealing with new weeds.

These considerations cast doubt on the first two arguments above. The last two arguments deserve closer examination. Again, they are arguments meant to persuade as well as inform, and so we must pay particular attention to their rhetorical strategies.

The first argument concerns the power of the chemical industry, an issue that Doyle has raised in his book, *Altered Harvest*. In the chapter titled "Magic Molecules, Clever Chemistry," Doyle recounts the history of the development of E. I. Du Pont de Nemours and Company, one of the world's larger multinational corporations and a leader in the race for GEHR crops.[92] Why have large conglomerate chemical companies suddenly become so enamored of seed companies? What will the giants do next?

They may want to extend the life of the chemical division's old agricultural chemicals. Industry seems as interested in *creating* needs for its chemicals as it is in meeting already existing needs. Du Pont, for example, will use herbicide resistance crops not only as a way to sell its new sulfonylurea herbicides, but as a way to breathe new life into its older compounds. It need only genetically engineer tomatoes and beans resistant to the old standbys to make them profitable once again.

In a footnote, Doyle expresses the objection in his own voice:

Although a chemical or pharmaceutical corporation may spend

> as much as $50 million . . . developing a new pesticide [or herbicide] . . . a popular patented substance, in a few years time, will produce annual revenues that may run as high as $500 million to $1 billion. Eli Lilly's herbicide Treflan, reaping $350 to $400 million annually between 1979 and 1981, has accounted for at least 10 percent of the company's total corporate income since 1978. Eli Lilly's herbicide Treflan, reaping $350 to $400 million annually . . . has held as much as 70 percent of the dinitroaniline herbicide market in recent years.

And how many competitors does Lilly have? Apparently, only two: "Similar herbicides from American Cyanamid and BASF hold the other 30 percent. . ." The footnote ends with the author summarizing the case by again quoting from a magazine: "'Products like Aatrex, Treflan and Roundup,' says *Chemical Week*, 'guarantee years of high earnings.'"

Markets in which a half dozen companies or less control 55 percent or more of the business are called shared monopolies. Many economists think shared monopolies are problematic because they allow companies to collude with one another, fixing prices at higher levels than the prices would be if there were true competition in the industry. In the dinitroaniline herbicide market, for example, three companies control the entire market. Is this good for those who need to buy dinitroaniline chemicals? Is it good for the economy generally?

If readers are to be convinced to oppose GEHR research on the grounds that shared monopolies in the chemical industry represent a threat to moral principles, however, we need a more subtle economic analysis than is provided in *Altered Harvest*.[93] Consider that another vocal critic of pesticides has complained that there is *too much* competition in the industry:

> The problem today [in 1978] is that there are too many companies with too many products battling for the swag; fourteen hundred pesticides and thirty thousand labels. What a joke! This forces the chemical companies into a merchandising dogfight and into continuously seeking another DDT or parathion; that is, a low-cost biocide designed more to capture markets than to fit into scientifically conceived, integrated pest-management systems.[94]

Before leaving the issue of industry power, let me summarize the case against *O* on this point. It is an open question whether GEHR crops will encourage a more concentrated biotech industry. Many startup biotech companies went out of business in the mid-1980s. This happened more because of internal business problems, like the failure to produce a money making discovery, or a lack of managerial expertise, than because of collusion among the giants. Even the giants have found that a strategy of joint research ventures and reciprocal licensing arrangements may prove in the long run to be a more advantageous strategy than leveraged buyouts aimed at monopolistic control of markets. This is evidence that stabilization and cooperation may explain more of what is happening in the industry than cutthroat competition and concentration.

Of course, one may reply that cooperation between companies serves concentration, and is bad for the efficiency and competitiveness of the industry. But this reply must take account of the international character of the biotech industry where companies of one country are competing against those of another. For example, seven U. S. companies in the superconductivity race announced a joint venture in June of 1989 in order to try to compete against the Japanese, who hold a clear lead in this capital-intensive research and development field. Ag biotech companies in the U. S. may be coming to see the wisdom of such an approach, an approach that serves the interests of international competition by encouraging intranational cooperation. Doyle's concerns about the power of the chemical industry in the United States are not wholly unjustified, but further evidence is needed to justify the claim that we should oppose GEHR research on these grounds.

The last *O* argument that I want to consider was the first one raised in this article, that genetic engineering of plant varieties is wrong because it crosses species boundaries placed in nature by God. If God made boundaries in the plant and animal kingdoms, who are we to violate them in our laboratories?

The problem with this argument is that it begs the question about what a "species" is. Like "weeds" and "safe," "species" is a context-specific concept. Its meaning depends on how it is being used. Viewed from the perspective of animal breeding, species are not fixed because normal members of two different species (horses and asses) can give rise to members of a different species altogether (mules). Viewed from the perspective of evolutionary genetics, species are not fixed because species can give rise to mutations which become the parents of a novel species, and because the differences between members of one species (e.g., the Norwegian elkhound and the chihuahua, both dogs) can be greater than the differences between members of two different species (e.g., the Norwegian elkhound and the wolfhound).[95] Viewed from the perspective of biology, species are not fixed because nature gives ample evidence of fluctuations, transformations, and generally fluid boundaries between species.

Proscribing the crossing of species borders seems to depend on the denial of all of these perspectives, a denial which is usually accompanied by particular religious beliefs. The bald assertion "species boundaries ought not to be crossed" begs the question of what "species" means. Without telling us exactly how a species is constituted in the broader pattern of relations between organisms, we have no way to tell whether rigid species borders actually exist, much less whether they should or should not be crossed. Further, the proscription seems to rest on a great deal more than scientific principles and, without sharing the religious convictions that support *O*, it is difficult to see how one could think it binding on others.

Whether we are morally justified in crossing plant species seems to me no more debatable in the 1990s than whether we are justified in killing weeds. Few in this country could hold to such a view without having to alter their diet dramatically. But if the question does not seem compelling in the arena of plant life, it becomes a different matter when we cross into the animal kingdom. Sentient beings who can experience pain and emotion have interests that may well be thwarted if they are the product of two species unrelated in nature. We will consider the matter of crossing species in the next chapter when we take up the issue of transgenic animals. But when it comes to nonsentient plants, there is little room for moral concern about mixing varieties.

Like *E*, *O* is based on dubious assertions, attacks on straw men, and circular argumentation. Those trying to develop an environmental perspective on the morality of agricultural biotechnology research may rest no easier with *O*'s blanket condemnation of publicly funded GEHR research than with E's carte blanche endorsement of it.

5.3. Qualified Endorsement (*QE*)

Consider first the view that GEHR research using public funds is *prima facie* permissible, that is, permissible unless and until it is shown that the research leads to products, practices, or attitudes that conflict with the environmental, economic, and communal principles stated earlier.

Those who would hold to a position of qualified endorsement might argue that the research is important in order to meet food demands and to promote a greater balance of good over evil, but would also acknowledge the validity of the questions I have raised about the technology.

Those in *QE* are likely to be most concerned with the rights of individuals, autonomy, reciprocity, and the likelihood of agricultural research to lead to a greater ratio of good over bad consequences. Someone with this view would like to know how GEHR research products will actually measure up to ethical principles before rendering a final judgment about it. In the meantime, they feel comfortable enough with the research to presume it innocent until proven guilty.

Those in *QE* would claim not that GEHR crops are safe, but that the crops are safe enough for this or that purpose given the current agricultural system and world food needs and the current implied or explicit social contracts on which that system is based. An environmentalist might hold to *QE* on the basis not only of the economic benefits already acknowledged but on the basis of potential ecological advantages; the new chemicals appear to be safer for us and the environment.[96] They do not leach into groundwater the way alachlor and atrazine do because they adhere tightly to the soil. In the soil, the chemicals may be broken down into harmless molecules such as carbon, nitrogen, and oxygen. In some soils, the half-life of glyphosate is less than a week and in most soils not longer than five or six months.[97] And by making land currently in production even more productive, GEHR technologies might allow some agricultural acres to be returned to wildlife habitat or even wilderness.

If we carefully hedge our claim with all of the qualifications introduced earlier, we may assert that the new chemicals--the glyphosates, the imidazolinones, the sulfonylureas--are safer, simpler, and more effective than the old chemicals. The research has other features to recommend it. It contributes to our knowledge of basic subcellular plant structures, plant metabolism, and the role of genetic information in plant growth regulation; advances in basic scientific understanding of these biochemical mechanisms might enable us to design crops in the future that would not need herbicide applications at all. In theory, one can envision tomato and corn plants so hardy that they could grow in the presence of weeds, or send out their own environmentally benign chemicals to kill just those weeds that actually compete with them. In such an ideal world, farmers would be freed altogether from their dependency on herbicidal chemicals.

What we need, those in *QE* might argue, is a balanced view of agricultural chemicals. So argues the president of the Connecticut Farm Bureau, Mary Potter:

> We have lost our national sense of balance when, because of faddist acceptance, toxic, dangerous, raw plant compounds [such as the abortion agent pennyroyal, the potentially fatal poke plant, and chamomile tea, which "can cause severe reaction in people who suffer from certain allergies"] can be offered to the public under the guise of health foods, while highly tested chemicals are branded as potentially unsafe.
>
> The suspension of the chemical weed killer 2,4,5,-T -- because of . . . potential involvement in miscarriages--

> ironically now allows the unchecked growth of hundreds of poisonous plant species with proven abortive abilities. Nettle . . . the root of a water hemlock plant . . . lupine . . . bracken fern . . . [can all result in abortion or] glaring birth defects in cattle [that feed on them].[98]

Potter seems to think that mere common sense will lead all of us to the same conclusion about herbicides. With the restoration of "a national sense of balance" we would all agree, she asserts, to repeal the Delaney provisions of the Food, Drug, and Cosmetic Act, requiring the banning of any chemical in which any detectable level of toxic materials is found. Adopting "the scientific approach" to "economics and the long-term public good" would cause us to bring back DDT, Mirex, cyclamates, DES, Red Dye Number 2, and saccharin, along with 2,4,5-T. Or so Potter's intuitions tell her.

Homer LeBaron, vice president of Ciba-Geigy Corporation, may not share Potter's particular judgments about the need for specific compounds, but in a speech given at Brigham Young University in 1988, "Ethics in the Agricultural Chemical Industry," he argued for a similarly "balanced" view of agricultural chemicals, a view based on what he called "reason, logic, objectivity and ethics."[99] Such a view stresses the fact that our food supply is much safer than most people think.

Let us consider a few of the slides LeBaron's used in his presentation. The first slide presents the conclusions of a study by Bruce Ames published in the journal *Science*.[100]

Slide 1

1. The incidence of specific kinds of cancer differs markedly in different societies.

2. Dietary factors are implicated to play a significant role in the incidence of some types.

3. No epidemiological evidence to suggest pesticide residues in food have contributed to increased cancer in U.S.

- Ames, Bruce N., et. al. 1987. Ranking Possible Carcinogenic Hazards. *Science* 236: 271-279.

Notice claim 3, concerning the lack of evidence linking pesticides to cancer. Does this claim contradict those cited above about nitrates in the groundwater and cancers in farmers? It is important to keep distinctions clear here. The problem with cancer among farmers stems, allegedly, from use of nitrogen fertilizers, not chemical herbicides. Nitrates in groundwater are a different problem from pesticide residues in or on food.

Notice, too, claim 4 from the same study, highlighted in the next slide:

Slide 2

4. Threat posed by natural "toxins" in plants is estimated to be at least 10,000 times greater than pesticides residues.

5. Cancer rates in US have remained relatively constant for the last 50 years, except lung cancer from smoking and melanomas from UV light.

Some foods grown organically present more of a cancer risk than foods grown with pesticides. Peanut butter, Ames points out, is one of these foods. If I feed my son Benjamin four tablespoons of peanut butter a day for the rest of his life, I will expose him to a not insignificant carcinogenic risk because peanut butter contains aflatoxin. It is several times more dangerous to eat a tablespoon of peanut butter than an apple grown with Alar, and there is a far greater carcinogenic hazard from drinking two cans of beer. But most of us would not consider the risk from peanut butter significant, nor deem it a threat to our children's well-being.

But considerations of risk need to be supplemented by considerations of informed consent. Risks we have freely assumed (such as an adult's choosing to eat peanut butter) are different from those imposed on us (such as a child's being given nothing but peanut butter to eat). The difference is important because many people do not know the risks associated with eating certain foods and might not choose to assume those risks if they knew what the risks were.

In order to protect the innocent, we authorize political and regulatory agencies to prohibit the sale of certain kinds of risky foods. If it seems silly to consider outlawing peanut butter because of what seem to be minute risks from aflatoxin, consider the fact that we currently prohibit the use of compounds that have *any* probability of causing cancer. Commenting on this rule, Henry Shue writes that

> Under current U. S. law . . . a substantial probability of harm is not now considered necessary to the case for regulation, and therefore a low probability is not sufficient to weaken the case [for regulation], when the degree of seriousness is maximal (as it is taken to be when the harm at risk is in fact cancer) and the risk is being inflicted upon some people by others coercively.[101]

So, while it would appear to be unjustified to worry about the natural carcinogens in the two tablespoons of peanut butter Benjamin eats each day, it would not be unjustified to worry about other sons whose families are constrained by tight budgets and who may be eating five or six tablespoons of peanut butter a day. If Ames' worries are scientifically justified, regulation of the sale of peanut butter on Shue's moral grounds could be defended.

LeBaron did not discuss these issues. Consider another slide:

Slide 3

Carcinogenic risks are very difficult to assess because:

1. Time delay between exposure and response.

2. Differential response of different species.

3. Little evidence to support the assumption that carcinogenic tests with other species are good indicators of effects on people.

Wilson, R., et. al. 1987. Risk Assessment and Comparisons. *Science* 236: 267-270.

Notice the last claim. Most of the chemicals banned in the US have never been proven to be carcinogenic to humans.[102] They have been proven to cause cancer to lab animals when administered in large doses over a long period. Whether the chemicals actually constitute a threat to humans is another matter.

LeBaron also projected this image on the screen:

Slide 4

HOW EXTRAORDINARY!

The richest, longest lived, best protected, most resourceful civilization with the highest degree of insight into its own technology is on its way to becoming the most frightened!

- A. Wildavsky, 1979

The rhetorical and political use of science in our society is an intriguing phenomenon. But how unobjective, emotional, and unbalanced is it to worry that your children might suffer from decreased immune function or develop serious allergy problems after forty years of being exposed to the synergistic and antagonistic effects of pesticide residues on beef, tomatoes, potatoes, oranges, and lettuce?[103] And how many people could articulate their fear in this way, so as to distinguish for an interviewer the difference between their fear of the chronic and the acute effects?

What is the actual level of risk from chronic exposure to pesticide residues on food? We do not know. In acknowledgment of this fact, a survey of top US scientists in 1987 caused the Environmental Protection Agency to rank this problem as one of the four most important issues it faces.[104] The four issues of "overall medium/high risk" identified in the report were:

1. "Criteria" air pollution from mobile and stationary sources (includes acid precipitation).
2. Stratospheric ozone depletion.
3. Pesticide residues in or on foods.
4. Run-off and air deposition of pesticides.[105]

Two of the EPA's top four priorities involved pesticides. The EPA is not worried about the risk of accidental death from pesticides (although they did express concern over the safety of workers who manufacture and apply them). Indeed, they concluded, that pesticides rank relatively low in cancer and noncancer health risks. They are concerned that pesticides carry high "ecological" and "welfare" risks stemming from point and nonpoint sources of surface water pollution and physical alteration of aquatic habitats (including estuaries and wetlands).[106]

The EPA's survey confirms the judgment of the insurance industry, that legal liability for deaths from pesticide poisoning are highly unlikely for most of us. Most college students and women voters probably should not fear this if they do.

But, again, we must look closely. Fears about sublethal, chronic health problems associated with pesticide residues in or on foods, and about environmental despoilation from pesticide pollution, *may* be justified.

The qualifications appropriate to interpreting this information are not only not provided but are actually buried under misleading talk about "actual risk." It is extremely difficult to figure out what is the "actual" level of the risk of your dying accidentally. The risk of sustaining a fatal injury while engaged in some activity is assessed with a significant range of error and refers to the dangers of a certain class (not individual) whose members typically engage in certain forms of behavior while restraining from others. The risk is assessed by a specific group for some purpose. The probable levels of accidental death for a US citizen who does not farm or work on a farm or chemical plant as assessed by the insurance industry for the purpose of establishing actuarial tables will be different from the probable levels of accidental death of a migrant California farm laborer.

If you are an airline pilot, the "actual" level of risk of your dying from "Commercial Aviation" is much higher because you fly often. If you have never flown and never will, the actual risk of your dying from commercial aviation is much lower. Risks, therefore, should not be described as "real" or "not real." They should be described as probabilities in the context of some specific web of human purposes, contexts, needs, and responsibilities.

But even if we granted all of these qualifications, would not science still give us a pretty good estimate of the risks associated with pesticides for most Americans? Science can give us a good idea of how the insurance industry at a particular moment in history views the chances of its having to pay an accidental death claim from pesticide poisoning. But that is a far different, and far weaker, claim than the one Potter would have us believe.

Scientists are not agreed that the risks of pesticide residues on food are small.[107] It would be easy at this point to discuss at length the views of any one of the many ecologists or entomologists who, like Robert van den Bosch, have lamented the role of pesticides and the power of its proponents.[108] But consider instead the views of one of agricultural biotechnology's staunchest defenders.

In an article which explicitly states as its intention the desire "*to convince you* that biotechnological research" is "*essential*" and "*to put into perspective* the concerns about [its] safety," Brill, vice president of the biotech firm Agracetus, argues that agricultural biotechnology is needed to find replacements for the chemicals now in use.[109] And yet Brill is apparently not as sanguine about the use of pesticides as Homer LeBaron is. He writes that

> Twenty percent of the farmers in Illinois, according to a recent study, have consulted a physician at least once with an ailment related to the use of pesticides. More and more data accumulate that show that pesticides get into the human food supply and . . . that at least some of these pesticides are potentially carcinogenic.[110]

If as unalarmed a scientist as Brill is concerned about pesticides, how strange is it that a fair number of other educated folk are too? What was once extraordinary now appears hardly puzzling at all.

There is an irony here. On the one hand, the chemical industry seems to want to assure us that there is little to worry about. On the other hand, it tells us that its biotech research wings are pursuing this line in order to find a new generation of safe chemicals. But which is it? Is the present generation safe or not? If it is safe, then we should not need yet another "new" generation. If it is not safe, then the industry should not be giving presentations that implicitly ridicule those who are concerned about its safety. Even the industry seems to have a divided mind on the matter of the safety of agricultural chemicals. Is it any wonder that the populace at large is confused?

Slide 3, quoting Wilson, stressed the importance of carefully qualifying risk assessments. But, in a passage LeBaron chose not to comment upon, Wilson further wrote that the *way* we compare risks is very important because it is so easy to mislead an audience. Risks "appear to be very different when expressed in different ways." One could argue, for example, that the Chernobyl disaster would produce as many as one hundred and thirty one *additional* cancers in the population of those in the plant's immediate vicinity, thus justifying the judgment that Chernobyl was indeed a "disaster." But using the same data and dividing the 131 cancers "by the approximately 5,000 cancer deaths expected in that population from other causes," you could also argue that Chernobyl would "*only* [produce] a 2.6 percent increase" in cancer cases, perhaps not a disaster at all.

Shift the context again and include all of the 75 million people in the Byelorussia and Ukrainian regions around Chernobyl and the result would be a rise in cancers of less than 0.005 percent, clearly an insignificant--if not negligible--increase.[111] On this calculation, Chernobyl is hardly worth thinking about. Wilson insists on the point. Failing to specify the frame of reference used in comparing risks or blurring the differences between competing frames of reference may lead to unwarranted conclusions and mislead one's audience.

These methodological points in Wilson's study are as important as the more specific conclusions listed in the slide LeBaron presented, but LeBaron chose not to mention them, focusing instead on Wilson's claim that there is little evidence to support the assumption that carcinogenic tests on animals tell us about the carcinogenic effects on people. Let's turn our attention to this claim.

Recall that the effect was to undermine one's confidence in the reliability of those studies which showed some ag chemicals to cause cancer in rats. Now, it is unclear whether LeBaron was directly quoting Wilson or merely summarizing his views. I do not find LeBaron's claim in Wilson's article. What I find in Wilson is this: "the comparison of carcinogenic potency in animal and man . . . require(s) a certain amount of theory."[112] Consequently, the step of extrapolating from animals to humans, Wilson admits, is "controversial." Nonetheless, he writes, such comparisons are useful. For example, animal studies showing that chloroform in drinking water is 20 times as likely to cause cancer in rats and mice as trichloroethylene seems a reasonable basis upon which to conclude that "although neither [chemical] is known to cause cancer in people, we might expect that chloroform would do so about 20 times as readily" as trichloroethylene. Wilson's position on the use of animal studies in risk assessment seems to me significantly different from the one imputed to him by LeBaron. Unfortunately, I doubt that many in LeBaron's audience will look up and read the actual studies to which LeBaron referred.

The slides based on the article by Bruce Ames et al. came from a study published in the same issue of *Science* as Wilson's. Ames' piece is titled "Ranking Possible Carcinogenic Hazards," and its second paragraph begins with this sentence: "Animal bioassays and in vitro studies are also providing clues as to which carcinogens and mutagens might be contributing to human cancer."[113] Ames and coauthors go on to add that extrapolating from animal carcinogenicity tests to humans is a "difficult" procedure and that "there is little sound scientific basis for (it)." But they immediately add,

> Nevertheless, to be prudent in regulatory policy, and in the absence of good human data (almost always the case), some *reliance on animal cancer tests is unavoidable*. The *best use of them should be made*, even though few, if any, of the main avoidable causes of human cancer have typically been the types of man-made chemicals that are being tested in animals.[114]

Ames' actual claims shed a different light on LeBaron's implication that animal carcinogenic studies cannot be trusted to tell us about pesticide safety.

Consider last the frame in which LeBaron's plea for balance reaches its climax. How extraordinary that the most rich, long lived, well protected people on earth are becoming the most frightened! The rhetorical structure of Wildavsky's exclamation in this slide begins with two words all in upper case letters followed by an exclamation mark. The viewer is told in unmistakable terms that the message to follow is of grave importance.

Wildavsky's sentence is an example of what classical rhetoricians called "asyndeton," the deliberate omission of conjunctions between a series of related clauses: "The richest, longest lived, best protected, most resourceful . . ." Aristotle, master analyst of rhetoric, noted that the use of asyndeton is:

> especially appropriate for the conclusion of a discourse, because there, perhaps more than in any other place . . . [the rhetorician wants] to produce the emotional reaction that can be stirred by . . . rhythm.[115]

Scientists stirring emotional reactions? It is difficult to imagine an attentive audience member in Utah not feeling attracted to LeBaron's side. The effect of the Wildavsky slide must have been emotional, even hyperbolic. And hyperbole it is, since Wildavsky exaggerates the achievements of our civilization, "the richest," "best protected," etc., for the purpose of emphasis. When strung together and presented before the antithetical climax, the series of exagerations heightens the drama. The concluding clause juxtaposes the opening clauses of the asyndeton with an idea that is their exact inverse: We pampered people are, ironically, also the world's *most frightened*!

Imagine the effect of seeing this slide in life-size proportions in a darkened room at the end of the previous series of slides. Few must have come away feeling frightened about pesticides. But is the rhetorical use to which science is being put here a good one? We need not think that all herbicides are as carcinogenic, teratogenic, or ontogenic as 2,4,5-T or alachlor to see the problems with LeBaron's presentation. Few if any of the herbicides currently in use carry significant risks of accidental death for those who do not handle them directly. But as we have seen, there is a great deal we do not know about the risks associated with the chronic or environmental effects of these chemicals working by themselves and in relationship with other chemicals. According to the EPA, we know "little about complex chemical transformations involving pollutants in the atmosphere or groundwater," and "little about the reactions of entire ecosystems (as opposed to single species) to environmental pollution."[116]

Even if Wildavsky and LeBaron are right that we are an overly frightened civilization, that supposed fact about us should not appear "extraordinary." The magnitude of the problems we face, including but not limited to nuclear holocaust, the greenhouse effect, ozone depletion, species extinction, and tropical rainforest despoilation, is justifiably frightening. The problem with LeBaron's presentation is the same as we saw in Mary Potter's. Both think that society in general has an unbalanced view of pesticides and that we are not using common sense. And yet whereas "common sense" refers to the sense of a community, wisdom invested in a large consensus of the people, there is no "common" sense among Americans in general or scientists in particular about the safety of pesticides. Both communities are sharply divided over the issue. So the problem cannot be settled by appealing to common sense.

The position recommended by Potter, President of the Connecticut Farm Bureau, and by the vice-president of Ciba-Geigy may be morally defensible, but not on the grounds of common sense or "science." It is the view of a specific group of farmers and scientists, a group whose own interests are not, as Potter admits, unrelated to the judgment asserted. While Potter and LeBaron make strong rhetorical appeals for "balance" and "objectivity," their own positions are marked by a clear bias; views matching theirs are called "scientific," or "balanced," or "common sensical" while those not matching theirs are subtly alleged to be impractical, "fads," "extraordinary."

QE is not utterly indefensible by my environmentalist lights, but it fails to make the case that current expenditures of public funds on genetically engineered herbicide resistant crops are necessary or that the economic and environmental benefits expected from GEHR crops outweigh the risks. Those who hold *QE* may indeed have environmentalist leanings. But when they argue for GEHR research they should not appeal to abstractions such as logic or science, nor assume they have a corner on common sense and objectivity.

5.4. Qualified Opposition (*QO*)

In light of the extensive discussion of the case for qualified endorsement of GEHR crops, the case for qualified opposition can be stated succinctly. There are several reasons of a consequentialist sort why GEHR technology does not seem likely to help us develop an agriculture consonant with the moral principles I have espoused. Those reasons include the likelihood that GEHR crops will lead to: a diminution in the diversity, integrity, and beauty of farm ecosystems; an impoverished rather than enhanced form of farm life; an ever more tightly linked and concentrated agricultural economy with even higher entry barriers; an increase in the use of herbicides; and an increased concentration of land, wealth, and power in the farm supply and food processing industries.[117]

These judgments must all be made tentatively and, for reasons suggested above, ought not to be asserted without acknowledging a large probability of error. Some environmental risks, for example, might be offset by other environmental benefits; other environmental risks might be outweighed by economic gains. Suppose that GEHR crops should prove to be the only way to keep up with the world's growing demand for food. Would we then want to oppose it? The potential economic benefits of this technology should not be weighed lightly. Might GEHR crops contribute to slightly lower food prices for consumers--a distinct benefit to the poor and disadvantaged? Might they lead to an improvement in the efficiency with which land and labor are used in farming? It seems likely that they would force some farmers out of business, but the dislocations are not expected to happen as quickly or traumatically as is expected with bGH. And do we want to deny the gains from improved profit margins to stakeholders in multinational companies successfully marketing the seeds and chemicals?

From my perspective, there are genuine benefits that may accompany GEHR crops, so blanket condemnation of the technology is unwarranted. But we must weigh the importance of a growing economy against other factors. One way to do this is to ask how urgently farmers need the new seeds and chemicals. Might farming be a sustainable and profitable business without GEHR crops? There are alternative methods of weed control. Intensively managed fields may be rotated so as to reduce the severity of weed infestations that plague monocultures; fields may be cultivated when needed and weeds killed mechanically; as a last resort, present generation chemicals may be handsprayed on specific spots when rotation and cultivation fail to offer sufficient control.[118] Each strategy widely employed would slow the speed with which herbicide resistant weed species are appearing. Each strategy is not only consistent with our three moral principles but with the agroecological advice of biologists Levins and Lewontin:

> An attempt to control pests should begin with an examination of the whole ecosystem in its heterogeneity, complexity, and change. This runs counter to the usual paradigm, reinforced by the division of labor in applied science, of isolating the smallest parts of problems and changing things one at a time.[119]

Why examine the whole ecosystem? Why oppose GEHR crops as one step in trying to change more than one thing at a time? Because we do not know the effects such crops will have on us or the environment twenty or thirty years hence just as no one knew the effects of the arsenicals or London purple or 2,4,5-T or parathion or atrazine or alachlor when they were introduced. But even if we did know that GEHR crops would have no ill-side effect for any sentient creatures, the herbicide treadmill would still be unsustainable.

As Susan George argues, if we were to try to feed the world "an American diet, using U. S. agricultural production technologies (assuming oil were the only energy source) all petroleum reserves would be exhausted within eleven years."[120] GEHR crops will probably increase the productivity of each unit of labor in agriculture, but will not necessarily increase the productivity of each unit of land. This is because GEHR crops will favor the large scale, capital-intensive, style of agriculture, a style which often is not as productive per acre as smaller scale, more management-intensive farms.

But would smaller scale, labor-intensive, farms be profitable? In some circumstances, where the climactic and soil conditions were favorable, perhaps. In other circumstances, probably not. Since we are envisioning an agriculture that cuts down not only on pesticides but on purchased nitrogen inputs as well, we must ask whether a farm that got its weed control from less chemical intensive strategies could also get its nitrogen needs from legumes and animal wastes rather than synthesized ureas.

Worldwide, legumes might be able to provide even more fertilizer for farmers now dependent on manufactured versions. They would have to make several changes at once, however, changing from grain monocultures to mixed animal-and-grain farming. Such changes are never easy, and probably would require that many farmers work longer and harder hours. Few are likely to make such dramatic changes unless they see some profit in it.

In mixed farming, farmers must know where the weeds grow and how quickly they are likely to spread. Some weeds "may be restricted to wet or dry soils, sandy or clay, rich or poor soils, grasslands or cultivated lands, open fields or shady places, acid, netural, or somewhat alkaline soils."[121] And then the farmer must decide not only what the weed problem is, where it is likely to go, when it is likely to go there, and how harmful it is likely to be. The farmer must also decide how to deal with it: prevent it, block its spread, eradicate it chemically, or reduce it merely to the level of economic injury?[122]

As this discussion suggests, traditional farmers require a broad range of very specific and localized knowledge. They must make judgments of a practical sort, combining the wisdom of past seasons with predictions about the likely course of the future. If such judgments were not already very difficult to make, they are compounded by the fact that pests may be invulnerable to pesticide treatments during different stages of their lives. The alfalfa weevil consumes most alfalfa while it is a fourth-instar larva, but other insects do it as adults. So conducting integrated pest management well requires that mixed farmers know not only pest and weed densities, but the age distribution of the pest population, densities of beneficial insect and weed populations, condition of the crop, and expected changes in temperature and moisture.[123]

There are good reasons to encourage more of our farmers to develop and hone local knowledge, and the reasons are of an agroecological as well as anthropological sort. To those who would object that moving toward rotations and mechanical cultivation is trying to turn back the clock, we might reply as G. K. Chesterton replied in another context:

> There is one metaphor of which the moderns are very fond: they are always saying, "You can't put the clock back." The simple and obvious answer is "You can." A clock, being a piece of human construction, can be restored by the human finger to any figure or hour. In the same way society, being a piece of human construction, can be reconstructed upon any plan that has ever existed."[124]

The wager here is that the goal of getting off the herbicide treadmill is worth "putting the clock back" a bit. We might find that in doing so we have not gone backward but forward, to a new farm, which values properly the wisdom invested in local knowledge of the land.

Another reason for qualified opposition has to do with our relations with the Third World. When we continue to market chemicals in developing countries banned in our own, we show little moral reciprocity. In 1978, Imperial Chemicals (ICI) exported paraquat to Costa Rica while BASF (as previously mentioned) was sending 2,4,5-T and Dow was sending 2,4-D. In 1979, 2,4-D was sold to Colombia by BASF, Celamerck, Ciba-Geigy, Dow, and Shell. These chemicals were either banned or restricted for sale in the United States at the time.[125] GEHR crops and chemicals are sure to be advertised in overseas markets, markets in which farmers may not always know what the experts know about the risks of the chemicals. To the response that such farmers are free not to buy the chemicals, we must ask whether this is so. Free choices are only free to the degree that they are informed.[126] At least one farmworker quoted above testifies that workers usually do not know the risks. If this is true, can farmers lacking such information be said to be "free" to choose such products?

This issue raises questions about the rights of the most vulnerable discussed in chapter two. As Henry Shue writes,

> What about minorities that are vulnerable because of some reason other than lack of information? What about the badly informed, the badly educated, the children, the infants, and the unborn? These groups cannot protect themselves by reading labels.[127]

For those whose moral perspective makes the most vulnerable humans "the measure of all things," so to speak, the preference of a farmer to choose GEHR crops will not weigh very heavily if it turns out that the herbicides used with those crops endanger the safety of children.[128]

There are also questions about the environmental suitability of technologies such as GEHR crops. How compatible with the changing cycles of nature and human trade will this technology be? As Levins and Lewontin put it:

> The high-technology monocultures [typical of chemical agriculture] increase the vulnerability of production to natural and economic fluctuations. The plant varieties developed for the green revolution give superior yields under optimal conditions of fertilizers, water, and pest management. They have been selected to put most of their energy into grain rather than vegetative parts, and *the resulting stout dwarf stems make it easier for weeds to outgrow them, making herbicide use mandatory.* The reduced

> root growth increases the plant's sensitivity to a shortage of water. Irrigation buffers the crop against the vagaries of rainfall but increases the farmers' sensitivity to the price of fuel. High-nitrogen fertilizers and the growth-stimulating effects of herbicides make the plants more vulnerable and attractive to insects . . . And monoculture removes diversity as one of the traditional hedges against uncertainty.[129]

To worries that the tightly-linked international system of export agriculture has been unfair to developing countries, some reply that green revolution technologies have provided struggling nations with a wide variety of jobs and businesses.[130]

A free market system is, all other things be equal, a good thing, as is a diverse array of businesses. But not all countries have been able to adapt to export agriculture as well as others. Where changes have been cataclysmic a "wide array" of agribusinesses may never have started, or the mechanisms for feeding people may not have been in place during the transition from a subsistence to a cash economy. The result may have been that the beginnings of an agribusiness system may have done little more than turn subsistence peasants into unemployed city dwellers and induce the remaining farmers to raze forests and plow up hillsides.[131]

Should all countries be encouraged or induced to jump headfirst into relying on cash crops as the way to sustain their economies or will doing so lead to uneven development and environmental disaster?[132] Have we really been fair with our trading partners when we "erect high barriers to imports of temperate-zone products from developing countries and then subsidize [our] own exports"?[133] Is it fair that world markets for ten major Third World exports are shared monopolies controlled by three to six multinational corporations?[134] By the standards of autonomy and reciprocity, we must be careful not to exercise more power than is warranted in our relations with farmers in less developed countries.[135]

For all of these reasons, those who share my economic, environmental, and communal principles will have a difficult time being convinced to approve of GEHR research. Public funds can be spent in much better ways: increased research on low input sustainable agriculture; attempts to map the carrying capacity of various geographic areas and determine the optimum pasture and crop land uses; more funds for extension and education of farmers trying to control weeds through cultural practices; economic studies of ways to encourage smaller farms on which labor (e.g., hand weeding and mechanical cultivation) may be more easily substituted for capital inputs such as herbicides; more research on biomass as an alternative source of energy and chemical feedstocks; better programs at universities in environmental studies, ecology, animal behavior, wildlife management, and evolutionary biology. [136]

The difficulties involved in studying these last areas make it all the more urgent that we attend to them. Sir Humphrey Davy once wrote "The larger the light, the larger the circle of darkness around it." Botanist G. Clifford Evans introduced his study of the problem of specialization in scientific knowledge by reminding us of Davy's vivid metaphor. "As our knowledge has grown," Evans explained, "so have the number and complexity of the unsolved problems, and many need highly specialized knowledge and techniques for their solution."[137] But even more require an interdisciplinary skill. Here, Evans repeated a comment of A. S. Watt's:

> Clearly it is one thing to study the plant [cell or even the] community and assess the effect of factors which obviously and directly influence it, and another to study the interrelations of all the components of the ecosystem with an equal equipment in all branches of knowledge concerned.[138]

Research efforts in environmental studies, ecology, and evolutionary biology may not be "ag biotech" projects according to our strict definition. But they are, from my perspective, projects more deserving of scarce public funds than GEHR research.

But even if one could change only the allocation of biotech funds within the area of molecular biology, other projects would seem to outrank the herbicide resistance. More important areas might include research to devise vetches and cover crops to blanket bare Illinois fields in winter, and crops such as Kentucky fescue that can inhibit the growth of weeds (such as trefoil) through allelochemical effects;[139] research to introduce chemical molecules into crops that may inhibit the growth of weeds by the release of toxins;[140] research to find cornplants that can fix their own nitrogen, thus reducing our reliance on purchased fertilizer inputs, or with their own internal defenses against the European cornborer;[141] beans, tomato, and cotton plants resistant to lepidopteran insects;[142] species that protect each other through allelopathic effects so as to inhibit the growth of pests and weeds through biological means; and so on.

Are GEHR crops compatible with sustainble agriculture? A very modest research effort in this area might be justified to the extent that it would help us to understand basic plant and herbicidal mechanisms, and to the extent that it might help weed scientists to keep up with changes in the makeup of the weed flora. But where net economic returns could justify cutting back on pesticide usage--on many farms in the corn belt and in the semiarid northwest, for example--we should encourage this strategy.[143] Alternatives to GEHR crops are available. With an infusion of low cost labor resources or governmental subsidies for low input agriculture, present yields might be sustained not only without GEHR crops and chemicals but even, perhaps, without any of the current herbicide mixtures. In other areas and in other crops, herbicides and fertilizers are, without a doubt, necessary in the short run to insure a stable supply of efficiently produced food without increasing the amount of environmentally sensitive land used in production. Chemical pesticides are by no means uniformly bad from my perspective.[144] But we have seen good reasons to try to lessen our dependence on them.

From my perspective, GEHR research using public funds is impermissible until it can be shown that the research will not lead to products, practices, or attitudes that conflict with environmental goals.

A word about the status of my judgment. I do not believe my preference for *QO* over *QE* can be justified by appeal to transhistorical standards. The judgment is justified by concrete historical considerations such as the particular memories and aspirations of historical communities. In my assessment of GEHR research, I have in mind what might be called a traditional ideal of what constitutes good farming, an ideal dependent upon the agricultural communities where it is still practiced. I believe that the web of relationships required to make GEHR crops successful is inconsistent with the narrative tradition my aunt and uncle strive to embody. More than any scientific or theoretical considerations, this particular, historically conditioned, judgment inclines me to *QO*.

But particular ideals of farming should not automatically bind everyone, because they depend on cultural norms and even religious convictions not universally shared. At least one community, mass American consumer society, seems to value a way of life that apparently requires large scale tightly knit agriculture. The ideal of good farming for that community may be antithetical to my aunt and uncle's ideal. How would we resolve a dispute between the American consumer's wishes about farms and my extended family's ideal?

It is impossible to appeal to yet a third ideal of farming in order to argue that one of the two ideals is superior to the other. There is no ahistorical Universal Ideal of Good Farming to tell us whose ideal of farming is the true one. We need not expect everyone to be bound by one person's notion of good farming. Nonetheless, we should realize the historically conditioned nature of all ideals of good farming, including the ideal of tightly linked modern agriculture.

It is from the perspective of my own interests and purposes that I offer the historically conditioned and qualified judgment that GEHR research is morally inappropriate.

- - - - - - - - - -

In his short story, "The Birthmark," Nathaniel Hawthorne tells of a scientist named Alymer whose deep love for his young bride, Georgiana, is matched only by a strange obsession to remove a birthmark from her cheek. There is no reason for Alymer's research because the woman is beautiful even with the supposed imperfection. And Alymer's quest requires that she be put through a series of painful experiments, the chronic effects of which are unknown. The tragedy is heightened when the narrator informs us that Georgiana's mark is not unlike those that "Nature, in one shape or another, stamps ineffaceably on all her productions."[145]

Nevertheless, Alymer is soon working night and day in his lab trying to find the chemical liquid that will perfect Georgiana's complexion. When he finds it at last, he gives it to her. Obedient to the end,

> She drinks it, and her birthmark disappears; she is perfect; but she no longer belongs to nature. She calls to her husband: "you have rejected the best thing the earth could offer. Alymer, dearest Alymer, I am dying."[146]

The moral of this story for ag biotech as been admirably drawn by Mark Sagoff:

> In [Alymer's] passion to make [Georgiana] perfect, he lost sight of the value of what he already possessed. We, too, are likely to succeed at many of the purposes to which we put recombinant DNA technology. But we must proceed with reflection and caution lest, in our passion for power and profits, we lose more than we gain by our success.[147]

Hawthorne's story anticipates Greg Brown's message in "Walkin' the Beans." Our obsession with how something *looks* may set us tasks we can never ever get done. With weeds, looks can be deceiving. Perhaps we need to retrain our eyes to see the beauty of selected weeds between rows of corn, and to see the beauty of wholistic approaches to weeds and farming.

If we cannot do this, we may lose more than we gain by our success in the ag biotech lab.

Notes

1. Greg Brown, "Walkin' the Beans," *Iowa Waltz* album.
2. Jeremy Rifkin, *Algeny: A New Word--A New World* (Harmondsworth: Penguin, 1983).
3. C. S. Lewis, *Mere Christianity* (New York: Macmillan, 1952).
4 The assertions I have just made about moral laws are contentious; not all moral philosophers are convinced that moral laws exist. To argue adequately for a theoretical position like metaphysical realism, however, would take us far afield. Rather than getting bogged down in arguments about moral ontology, I have meant only to present my own views.
5. Richard Levins and Richard Lewontin, *The Dialectical Biologist* (Cambridge: Harvard University Press, 1985), p. 271.
6. For two anthropological works on this subject, see Mary Douglas and A. Wildavsky, *Risk and Culture* (Berkeley: University of California Press, 1982); and E. Larson, *Food: Past, Present, and Future* (London: Frederick Muller, 1977).
7. Harmon Henkin, Martin Merta, James Staples, *The Environment, the Establishment and the Law* (Boston: Houghton Mifflin, 1971), pp. 123-124.
8 Nor have such cultural practices disappeared entirely from American agriculture; a five year rotation is currently advocated by the Iowa based organization, Practical Farmers of America, in which corn, soybeans, corn, oats, and hay are grown in successive years. All of the crops are fed to livestock, and these are marketed at appropriate times.
9. James Whorton, *Before "Silent Spring:" Pesticides and Public Health in Pre-DDT America* (Princeton: Princeton University Press, 1974), p. 5.
10. Whorton, p. 18. The Colorado potato beetle on Long Island has now developed genetic resistance so that no registered pesticide can control it.
19. E. F. Adler, W. L. Wright, and G. C. Klingman, "Development of the American Herbicide Industry," in Jack R. Plimmer, *Pesticide Chemistry in the 20th Century* (Washington, D.C.: American Chemical Society, 1977), p. 41.
12. Freeman L. McEwen and G. R. Stephenson, *The Use and Significance of Pesticides in the Environment* (New York: John Wiley & Sons, 1979), p. 110.
13. Cf. McEwen and Stephenson, p. 112.
14. McEwen and Stephenson, pp. 111-112.
15. Whorton, p. 181-2. Cf. Robert L. Rudd, *Pesticides and the Living Landscape* (Madison: University of Wisconsin Press, 1964), pp. 15-16.
16. Sterling Brubaker, *To Live on Earth: Man and His Environment in Perspective* (Baltimore: Johns Hopkins University Press, 1972), pp. 79-80.
17. Whorton, p. 25.
18. *Practical Entomologist* I, 4 (1865), cited by Whorton, p. 17.
19. Cited by Whorton, p. 68.
20. Whorton, p. 26.
21. DDT was banned for use in the US in 1973. It is still used in many developing countries to control lice, typhus, malaria and other insect-born diseases. In part, the usefulness of DDT in these countries is based on the fact that the safer organophosphate and carbamate insecticides are too expensive. Cf. Brubaker, pp. 12 and 18; and G. T. Brooks, "Chlorinated Insecticides: Retrospect and Prospect," in Plimmer, ed. (1977).

22. Donald G. Crosby, "The Environmental Chemistry of Herbicides," in Plimmer, ed., p. 106.
23. McEwen and Stephenson, p. 2.
24. Adler, et al., in Plimmer, ed. (1977), p. 39.
25. Adler, et al., pp. 41-43.
26. Kearney, in Plimmer, ed., p. 41.
27. Whorton, pp. 91-92.
28. Barry Commoner, *The Closing Circle: Nature, Man and Technology* (New York: Alfred A. Knopf, 1971), p. 149. Commoner points out that during this period the "population grew by 34 percent . . . [so that] crop production *per capita* increased 6 percent . . ." When one considers that the amount of acreage in production during this period *declined* 16 percent, the overall effectiveness of the nitrogen "declined fivefold" (p. 150).
29. Atrazine followed in 1958, and alachlor in 1969.
30. G. E. Barnsley, "The Future of Pesticides: Problems and Opportunities," in N. R. McFarlane, ed. *Herbicides and Fungicides: Factors Affecting Their Activity* (London: The Chemica. Barnsley identifies himself with Ciba-Geigy Canada.l Society, 1977), p. 1.
31. Levins and Lewontin, p. 212.
32. Kearney, in Plimmer, p. 37.
33. William R. Furtick, "Weeds and World Food Production," in David Pimentel, ed., *World Food, Pest Losses, and the Environment* (Boulder, CO: Westview, 1978): p. 60.
34. Adler, et al., p. 42.
35. Adler, et al., p. 49.
36. Levin and Lewontin, p. 236.
37. Charles Perrow, *Normal Accidents* (New York: Basic Books, 1984). Cited by Joseph Rouse, *Knowledge and Power: Toward a Political Philosophy of Science* (Ithaca: Cornell University Press, 1987): 230. Perrow introduced these terms, Rouse explains, "to illuminate the occurrence of what he calls `normal accidents' in certain high-risk technological systems. These are accidents that are due not so much to the malfunction of a single component of a system as to multiple failures whose combination was not anticipated. He claims that such accidents are to be expected in systems that are complex and tightly coupled" (p. 230).
38. Stan G. Daberkow and Katherine H. Reichelderfer, "Low-Input Agriculture: Trends, Goals, and Prospects for Input Use," *American Journal of Agricultural Economics* (Dec 1989): 1160.
39. For relative prices of herbicides, see Benbrook and Moses, p. 58.
40. "2,4-D increases insect and pathogen pests on corn . . . In 1974 field tests . . . corn leaf aphid populations numbered 3116 per tassel compared with only 1420 per tassel in an untreated corn field . . . European corn borer attacks on 2,4-D exposed plants were significantly greater (70%) than on untreated plants (63%) . . . 2,4-D corn had more southern corn leaf blight lesions and significantly larger corn smut galls. David Pimentel, (1978): 180. Cf. David Pimentel, "Down on the Farm: Genetic Engineering Meets Ecology," *Technology Review* (Jan 1987): 28.
41. This figure includes losses to insects, molds, fungi and plant diseases such as blight. J. L. Apple, in Pimentel (1978), p. 41.
42. Daberkow and Reichelderfer, p. 1160, citing R. Olson, K. Frank, P. Grabouski, and G. Rehm, "Economic and Agronomic Impacts of Varied Philosophies of Soil Testing," *Agronomy Journal* 87 (1987): 492-99.
43. Moshe J. Pinthus, Yaacov Eshel, and Yalon Shchori, "Field and Vegetable Crop Mutants with Increased Resistance to Herbicides," *Science* 177 (25 August 1972): 715-716.

44. Charles M. Benbrook and Phyllis B. Moses, "Engineering Crops to Resist Herbicides," *Technology Review* (November-December 1986): 57.
45. Cf. JoAnne J. Fillatti, John Kiser, Ronald Rose, and Luca Comai, "Efficient Transfer of a Glyphosate Tolerance Gene into Tomato Using a Binary *Agrobacterium Tumefaciens* Vector," *Bio/Technology* 5 (July 1987): 726-730.
46. Cf. Tauer and Love, p. 1; Anonymous, *WSSA Newsletter* [Weed Science Society of America] 16 (1988): 8; Charles M. Benbrook and Phyllis B. Moses, (1986), p. 57.
47. Kai Nielsen calls these "moral truisms," and cites torturing the innocent and breaking promises as examples. See Nielsen, "Searching for an Emancipatory Perspective: Wide Reflective Equilibrium and the Hermeneutical Circle," in Evan Simpson, ed., *Anti-Foundationalism and Practical Reasoning*: *Conversations Between Hermeneutics and Analysis* (Edmonton: Academic Printing, 1987): 147.
48. Aldo Leopold, *A Sand County Almanac* (New York: Oxford University Press, 1949), pp. 224-225.
49. For different perspectives on the issue, cf. the views of Paul Taylor, *Respect for Nature* (Princeton: Princeton University Press, 1987); J. Baird Callicott, *In Defense of the Land Ethic* (New York: SUNY Press, 1989); and Tom Regan, *The Case for Animal Rights* (Berkeley: University of California Press, 1983), p. 372.
50. These hard and fast principles make morality seem like an open and shut case, but the moral problems that engage our attention are not easy. Typically, they are difficult, messy, complex, and indeterminate. To deal properly with them, we need a supple moral vocabulary, stocked not with the language of principles alone, but with language about virtues and vices, convictions and hunches, justice and caring, forgiveness and stubborness.
51. The National Commission for the Protection of Human Subjects of Biomedical and Behavioral Research issued The Belmont Report in 1979. This report contains ethical principles designed to protect humans used in scientific research. One of the principles by which research projects are to be assessed is whether they are likely to lead to a greater ratio of benefits over harms. According to Ruth Macklin of the Albert Einstein School of Medicine, however, research proposals are rarely rejected on the basis that they fail to promise utilitarian benefits, and this is because the *potential* good of most basic scientific research is literally inestimable, even if much basic scientific research never leads to any useful products. There have been too many cases where research led, unpredictably, to something like penicillin. Macklin lecture, "Human Subjects Research Today," in the "Social Impacts of Biotechnology" seminar, Biology Department, Princeton University, 28 February 1990.
52. I hope, of course, that my analysis will have some bearing on private GEHR research, but when I use the phrase "GEHR research," I shall mean publicly-funded research.
53. E is rarely articulated as an argument, appearing more often as an assumption undergirding such texts as explicit advertisements as well as purportedly objective journalistic articles in farm magazines, industry pamphlets, and highly regarded scientific journals. I have found E expressed in diverse sources, including such things as Monsanto advertisements in the journal *Bio/Technology* featuring a large full-color picture of what is obviously a "family" farm and in the well-respected professional journal *Science*: Cf. Winston Brill, "Safety Concerns and Genetic Engineering in Agriculture," *Science* 227: 381-384. To the extent that E is not argued for, it functions as a prejudice of the modern research establishment.
54. Benbrook and Moses, p. 58.
55. Benbrook and Moses, p. 59.

56. I follow standard practice in using "pesticides" as a generic term to include each of the following species of chemical pesticide: "insecticide" for killing insects, "herbicide" for killing weeds, and "fungicide" for killing plant diseases.
57. Adler, et al., pp. 41-43; Tauer and Love, p. 1.
58. Cf. Tauer and Love, pp. 10-12.
59. Phyllis B. Moses and Charles E. Hess, "Getting Biotech into the Field," *Issues in Science and Technology* (Fall 1987): 35.
60. Cf. my conclusion in Comstock (1987).
61. Benbrook and Moses, p. 61.
62. Peter Isacson, director of Epidemiology Division, University of Iowa College of Medicine. Quoted in Carol Rose, "Iowa's Tainted Water Hurts Unborn Babies, U of I Study Suggests," *Des Moines Register* 3 October 1989, pp. 1, 9a.
63. Benbrook and Moses, p. 56.
64. Benbrook and Moses, p. 60.
65. The estimate of the relative strength of TCDD and arsenic is that of Al Young, who is identified as a physiologist who testified in the Agent Orange trials, in Judith Cook and Chris Kaufman, *Portrait of a Poison*: *The 2,4,5-T Story* (London: Pluto Press, 1982): 17. On the teratogenic effects of TCDD, see Shane S. Que Hee and Ronald Sutherland, *The Phenoxyalkanoic Herbicides, Vol. I*: *Chemistry, Analysis, and Environmental Pollution* (Boca Raton: CRC Press, 1981), p. 262.
66. Cook and Kaufman, 15.
67. Hee and Sutherland, p. 262. The authors add this explanatory sentence: "In spite of the suggestive nature of the above, it does not prove cause and effect. However, the segment of the population most likely to bear the brunt of the spraying was also the one most likely to be underrepresented in the above statistics." They also discuss three other cases of "unintentional and uncontrolled" TCDD contamination, in northwestern Florida in the 1960s, eastern Missouri (1971), and Seveso, Italy (1976).
68. Cook and Kaufman, p. 19.
69. BASF exported the chemical to Costa Rica in 1978.
70. For example, researchers working with a well understood strain of algae discovered that the sole site of action of the sulfonylureas and of at least some of the imidazolinones is the actolactate synthase (ALS). See Tom Winder and Martin H. Spalding, "Imazaquin and Chlorsulfuron Resistance and Cross Resistance in Mutants of *Chlamydomonas rheinhardtii*," *Molecular Gen Genet* 213 (1988): 394-399. Glyphosate works by inhibiting EPSP synthase. This is a "mid-pathway enzyme of aromatic acid biosynthesis confined to prokaryotes, lower eukaryotes, and higher plants." John P. Quinn, Joseph M. M. Peden, and R. Elaine Dick, "Glyphosate Tolerance and Utilization by the Microflora of Soils Treated with the Herbicide," *Applied Microbiology and Biotechnology* 29 (1988): 511. Cf. D. M. Shah, et al., "Engineering Herbicide Tolerance in Plants," *Science* 233 (1986): 478-481.
71. Benbrook and Moses, pp. 56-57.
72. Benbrook and Moses, p. 57.
73. Steve Radosevich alerted me to some of these problems with Benbrook and Moses' claims.
74. Winston Brill, "Why Engineered Organisms Are Safe," *Issues in Science and Technology* (Spring 1988): 44-50.
75. Brill, p. 45.
76. Brill, p. 46.
77. Brill, p. 91.
78. Brill, p. 47.

79. Brill, p. 47.
80. Benbrook and Moses give the following reason for this assertion: forty years of experience provides very few examples of herbicide resistance genes being transferred from crops to weeds. The incidence of such transfers will not rise with GEHR crops. But Dekker differs, suggesting that rapeseed, being very close to wild mustard, might transfer its resistance. Moreover, "crops like amaranth, proso millet, or sorghum have weedy counterparts with which they could cross-pollinate." Dekker, p. 6.
81. Regal, of the University of Minnesota, is quoted to this effect in "Gene Engineers `Should Study Ecology,'" *New Scientist* 121 (4 March 1989): 23.
82. Moses and Hess, p. 40.
83. Benbrook and Moses, n.d., p. 25.
84. Cf. Ruth E. Galloway and Lauren J. Mets, "Atrazine, Bromacil, and Diuron Resistance in *Chlamydomonas*," *Plant Physiology* (1984): 469-474; and Guy della-Cioppa, et al., "Targeting a Herbicide-Resistant Enzyme from *Escherichia Coli* to Chloroplasts of Higher Plants," *Bio/Technology* 5 (June 1987): 579-584.
85. "Technologies and Market Forces Shape the Form of Agribiotech Products," *Genetic Engineering News* (February 1987): 16-19. Quoted in Mark Sagoff, "Biotechnology and the Environment: What is at Risk?" *Agriculture and Human Values* 5 (Summer 1988): 29.
86. Jack Dekker, "Ethical and Environmental Considerations in the Release of Herbicide Resistant Crops in Agroecosystems," unpublished manuscript, Iowa Agricultural and Home Economics Experiment Station, p. 6. Quoted with permission.
87. For a criticism of Brill's approach that faults him for not taking up questions relating to "equity issues and holistic concerns for rural communities or small ecosystems," see Rachelle D. Hollander, "Values and Making Decisions about Agricultural Research," *Agriculture and Human Values* 3 (Summer 1986): 37.
88. Sandra Postel, "Controlling Toxic Chemicals," in Lester Brown, ed., *State of the World, 1988* (New York: W. W. Norton, 1988), p. 121. Postel refers to Foo Gaik Sim, *The Pesticide Poisoning Report* (Penang, Malaysia: International Organization of Consumers Unions, 1985).
89. At a conference on Agricultural Biotechnology and Sustainable Agriculture at Iowa State University in the spring of 1989, industry representatives were sharply divided on this question.
90. DuPont's sulfonylurea Glean can work effectively in some circumstances in quantities that are one "one-hundredth that of other herbicides." Doyle, p. 211.
91. LeBaron, *op. cit.*, p. 14, Slides 27, "Rates of Application: Insecticides (1940-1980)," and 28, "Rate of Application: Herbicides, (1940-1980), Dicot control in cereals."
92. David Pimentel, (1987), p. 28.
93. For a similar analysis of the question of competition in the farm sectors, see the twin articles by Luther Tweeten and Bruce Marion addressing the question "Is the Family Farmer Being Squeezed Out of Business by Monopolies?" in Comstock, ed., *Is There a Conspiracy Against Family Farmers*? USF Monographs in Religion and Public Policy # 5, Religious Studies Department, University of South Florida, 1990. See, too, Clifton B. Luttrell, *The High Cost of Farm Welfare* (Washington, D.C.: Cato Institute, 1989).
94. Robert van den Bosch, *The Pesticide Conspiracy* (Garden City: Doubleday, 1978), pp. 204-205.
95. Cf. Andrew Goude, *The Human Impact: Man's Role in Environmental Change* (Cambridge: MIT Press, 1981), p. 67.
96. According to Floyd M. Ashton, "most herbicides have a relatively low toxicity to mammals including man." See Ashton, "Persistence and Biodegradation of Herbicides," Fumi

Matsumura and C. R. Krishna Murti, eds., *Biodegradation of Pesticides* (New York: Plenum, 1982), p. 118. According to the Environmental Protection Agency, the most toxic herbicide is dinoseb. Moderately toxic are: paraquat, bromoxynil, diquat, 2,4,5,-T and 2,4-D. Most other herbicides, Ashton writes, "present little hazard" (p. 120), and he concludes that "in general long-term herbicidal residues have not been a problem in environmental pollution" (p. 127), and there is not yet evidence of bioaccumulation of herbicides although this potential exists "and should be subjected to further study" (p. 128).

97. Table 9.1, "Laboratory experiments on degradation in agricultural soils of 14C-labelled glyphosate," in L. Torstensson, "Behaviour of glyphosate in soils and its degradation," E. Grossbard and D. Atkinson, eds., *The Herbicide Glyphosate* (London: Butterworths, 1985): 146, gives estimates as low as 3 days for the half-life of glyphosate in silt loam and as high as 22.8 years in sandy loam. Their conclusion is that "present knowledge of glyphosate behaviour and degradation gives no reason to suppose that the herbicide may cause any unexpected damage after application to the soil or elsewhere in the environment" (149). This is because their survey of over 30 studies showed that the chemical is "rapidly adsorbed on soil," "practically immobile," and degraded principally into the substance known as aminomethylphosphonic acid (AMPA) which "is also biologically degradable" (147-8).

98. Mary Porter, "A Plea for Balance," in *Scientific Dispute Resolution Conference on 2,4,5-T* (Park Ridge, IL: American Farm Bureau Federation, 1979): 5.

99. Homer M. LeBaron, "Ethics in the Agricultural Chemical Industry," unpublished paper presented at Agriculture Week Symposium, Brigham Young University, March 24, 1988, p. 3. Quoted with permission of the author.

100 Bruce Ames, Renae Magaw, and Lois Swirsky Gold, "Ranking Possible Carcinogenic Hazards," *Science* 236 (1987): 271-279.

101. Henry Shue, "Food Additives and `Minority Rights': Carcinogens and Children," *Agriculture and Human Values* (Winter-Spring 1986): 196.

102. Richard Wilson and E. A. C. Crouch, "Risk Assessment and Comparisons: An Introduction," *Science* 236 (17 April 1987): 267-270.

103. Wilson and Crouch observe that "there have been few attempts to perform risk assessments for biological end points other than cancer," but that we know, for example, "that pollutants in cigarette smoke cause at least as many deaths through heart problems as by cancer" (p. 269). We need broader ranging studies of the non-carcinogenic health hazards posed by pesticides.

104. U. S. Environmental Protection Agency, Office of Policy Analysis, *Unfinished Business: A Comparative Assessment of Environmental Problems* (Washington, D.C.: EPA, February 1987). Cited in Richard Morgenstern and Stuart Sessions, "EPA's *Unfinished Business*," *Environment* 30 (July/August 1988): 17.

105. Ibid., p. 17.

106. Ibid., p. 36.

107. The disagreement was clear at the 192nd national meeting of the American Chemical Society in September of 1986. Thomas K. Jukes of the University of California at Berkeley claimed that the media was supporting "a determined effort to destroy by misinformation the U. S. biotechnology industry." By contrast, Daniel S. Greenberg, editor of *Science and Government Report*, said "it is naive to assume that public confidence will result from better understanding of technical detail. The public is scared because there *is* something to be scared about." Reported in "Chemicals: Must Everyone Be Scared?" *Technology Review* 89 (Nov/Dec 1986): 2.

108. Cf. Robert van den Bosch, *The Pesticide Conspiracy* (New York: Doubleday, 1978).

109. Winston Brill, "The Impact of Biotechnology and the Future of Agriculture," in Kevin B. Byrne, ed. *Responsible Science: The Impact of Technology on Society* (San Francisco: Harper & Row, 1986): 72.
110. Brill, p. 85.
111. Wilson and Crouch (1987): 270.
112. *Ibid.*, p. 268.
113. Bruce N. Ames, Renae Magaw, Lois Swirsky Gold, "Ranking Possible Carcinogenic Hazards," *Science* 236 (17 April 1989): 271.
114. Ames, *ibid.*, p. 271.
115. Edward P. J. Corbett, *Classical Rhetoric for the Modern Student* (New York: Oxford University Press, 1965): 433.
116. Morgenstern and Sessions, *op. cit.*, p. 16.
117. Ruth Hubbard's judgment that increased production of insulin would lead to more insulin use: "If we produce more insulin, more insulin will be used, whether diabetics need it or not." Quoted by Krimsky, in Zilinskas, p. 239, n. 17.
118. David Pimentel, "Agroecology and Economics," in M. Kogen, ed., *Ecological Theory and Integrated Pest Management* (NY: Wiley), pp. 299-319.
119. Levins and Lewontin, p. 220.
120. Susan George, *Ill Fares the Land: Essays on Food, Hunger, and Power* (Washington, D.C.: Institute for Policy Studies, 1984), p. 11.
121. Muenscher, p. 49.
122. The economic injury level is "the lowest pest [or weed] density for which an insecticide [or herbicide] treatment is economically justified. By definition, if the pest density exceeds the economic injury level, the cost of an insecticide treatment must be less than the value of the difference between the yields expected with and without treatment. If the pest density is below the economic injury level, an insecticide applicaiton costsw more than it can be expected to save by preventing crop losses." The economic threshold is "the density at which control measures should be applied [to a crop] to prevent an increasing pest population from reaching the economic injury level." Christine A. Shoemaker, "Pest Management Models of Crop Ecosystems," in Charles A. S. Hall and John W. Day, Jr., eds., *Ecosystem Modeling in Theory and Practice* (NY: John Wiley, 1977): 547.
123. Shoemaker, p. 551. To suggest how difficult such judgments are, consider the formula Shoemaker gives "to predict the number of insects in a population at any given life stage at any time":

$$N_i (t + 1) = s_i (v,T)\, N_i (t)\, (1 - a_i(T)) + s_i(v,T)\, N_{i-1} (t)\, a_{i-i}(T).$$

124. Quoted in Martin Marty's *Context* 21 (June 15, 1989): 2. Marty writes that he found the quotation in *American Scholar*.
125. David Weir and Mark Shapiro, *Circle of Poison: Pesticides and People in a Hungry World* (San Francisco: Institute for Food and Development Policy, 1981), p. 77. But cf., Robert L. Metcalf, who claims that 2,4,5-T was not restricted until 1979: "Benefit/Risk Considerations in the Use of Pesticides," *Agriculture and Human Values* 4 (Fall 1987): 21.
126. Cf. Arthur Caplan, "The Ethics of Uncertainty: The Regulation of Food Safety in the United States," *Agriculture and Human Values* 3 (Winter-Spring 1986): 186.
127. Shue (1986): 198.
128. Shue (1986): 200.
129. Levins and Lewontin, pp. 235-236.
130. Martin Kreisberg, "Miracle Seeds and Market Economies," *Columbia Journal of World Business* (March/April 1969). Cited by George, *op. cit.*, pp. 29-30.

131. Cf. Frances Moore Lappe and Joseph Collins, *Food First: Beyond the Myth of Scarcity* (New York: Ballantine, 1978); James O'Connor, "Uneven and Combined Development and Ecological Crisis: A Theoretical Introduction," *Race & Class* 30 (1989): 1-11; Piers M. Blaikie, *the Political Economy of Soil Erosion in Developing Countries* (London: Longman, 1985); Amartya Sen, *Poverty and Famines: An Essay on Entitlement and Deprivation* (Oxford: Clarendon Press, 1981).
132. On the question of technology transfer and the appropriateness of "miracle seeds" for countries lacking abundant rainfall or access to capital see the work of Kenneth Dahlberg, "Ethical and Value Issues in International Agricultural Research," *Agriculture and Human Values* 5 (Winter-Spring 1988): 101-111.
133. This is a question raised by the World Bank's 1986 *World Development Report*, p. 11. Quoted in Danaher, (1989): 42.
134. Wheat, corn, coffee, cocoa, tea, pineapples, bananas, sugar, cotton, and forest products are all commodities in which 60 to 90 percent of the market is controlled by between three and six corporations, according to the UNCTAD Statistical Pocket Book, p. 39. Cited in Danaher, p. 43.
135. In the early 1960s, Shell and Dow were informed that DBCP, a soil fumigant, had been linked with reproductive disorders in animals. "Not until 1977, after thousands of reports of workers becoming sterile, or nearly so, did the US ban domestic uses." Norris, p. 20.
136. The idea here is to base the chemical industry "on indigenous renewable raw materials." See B. O. Palsson, S. Fathi-Afshar, D. F. Rudd, E. N. Lightfoot, "Biomass as a Source of Chemical Feedstocks: An Economic Evaluation," *Science* 213 (31 July 1981).
137. G. Clifford Evans, "A Sack of Uncut Diamonds: The Study of Ecosystems and the Future Resources of Mankind," *The Journal of Animal Ecology* 45 (February 1976): 31.
138. A. S. Watt, "Pattern and Process in the Plant Community," *Journal of Ecology* 35 (1947): 1-22. Quoted by Evans, p. 31.
139. Cf. Elroy L. Rice, *Pest Control: Allelochemicals and Pheromones in Gardening and Agriculture* (Norman: Oklahoma University Press, 1983), p. 42.
140. William E. Fry, noting that several "aggressive perennial weeds including quackgrass, Canada thistle, Johnson grass and yellow nutsedge can inhibit the growth of certain crops" suggests that "some crop accessions may also inhibit weeds. Effective weed suppressors have been found in collections of oats, wheat, and soybeans." Commenting on the work of other researchers, he adds that "A. R. Putman and W. P. Duke of Michigan State University, East Lansing, hypothesized that many presently cultivated species may have possessed chemical molecules capable of inhibiting weeds when growing in wild habitat. Such traits could have been lost through domestication with intense breeding and selection for specific desirable traits. It may be possible to reintroduce such chemical traits into at least some of our present crop species." Fry, "Plant Microbes: Beneficial and Detrimental," in John J. Crowley, ed., *Research for Tomorrow: 1986 Yearbook of Agriculture* (Washington: USDA, 1986), p. 142.
141. Crop Genetics of Hanover Maryland announced plans in December of 1988 to fieldtest genetically engineered sweet corn resistant to the corn borer. "EQB Halts Biotech Release," *The Ag Bioethics Forum* 1 (April 1988): 4. Doyle quotes the OTA as saying in 1979 that in the California vegetable industry as a whole "breeding for resistance to pests and diseases as a primary means of pest control has never received the recognition and funding that it deserves" (*Altered Harvest*, p. 189, citing Congress of the United States, Office of Technology Assessment, *Pest Management Strategies in Crop Protection*, Volumes I and II, October 1979).

142. Cf. David A. Fischkoff, et al., "Insect Tolerant Transgenic Tomato Plants," *Bio/Technology* 5 (August 1987): 807-813.
143. Net economic returns on some chemical farms in these areas were no higher than comparable farms not using chemicals. Cf. T. Cacek and L. Langer, "The Economic Implications of Organic Farming," *American Journal of Alternative Agriculture* 1 (1986): 25-29, cited in Daberkow and Reichelderfer, (1989): 1159.
144. Bruce Ames' studies of the relative carcinogenic risks of natural and chemical risks supports my perspective. For example, the main fumigant EDB was banned in the US because residues in grain could cause 3 cases of cancer in 1000 people. According to Ames' HERP index, EDB has a rating of 0.0004 percent. But the rating of tap water is 0.001 percent, and peanut butter is 0.03. So the risk of getting cancer from EDB residues in food is significantly less than getting it from the chloroform in chlorinated water or the aflatoxin in peanut butter. Moreover, the alternatives to EDB use, such as food irradiation, or increased levels of mold in grain, may also be worse than the cure. So our attitudes toward chemical pesticides need, again, to take into consideration the widest possible range of relevant factors. See Ames, pp. 273, 277. Also, cf. Keith C. Barrons, "The Positive Side of Pesticides," National Council for Environmental Balance, Inc. 4169 Westport Road, Box 7732, Louisville, Kentucky, 1988.
145. Thus Mark Sagoff's account of the narrative. I am indebted to an article of his both for calling my attention to Hawthorne's story and for seeing how it illuminates our fascination with biotechnology. Mark Sagoff, "Biotechnology and the End of Medicine," *APA Newsletters on Computer Use in Philosophy . . . Philosophy and Medicine . . .* 88 (March 1989): 85.
146. Sagoff, p. 88
147. Sagoff, p. 88.

Chapter 3

Against Transgenic Animals (1992)

> *It is not by mere chance that Virtue . . . dwells in greatest proportions precisely upon that same span of soil where hogs thrive in greatest abundance. In Iowa, where people . . . read the Bible in the bathtub, there is approximately a full litter of pigs . . . for every single citizen.*
>
> -William Hedgepeth [1]

When I wrote "The Case Against bGH" in the late 1980s, I enjoyed eating meat, enjoyed serving it to my family, and believed one could simultaneously defend traditional family farms and the welfare of animals. Shortly after finishing that article, I read again, and more carefully, Tom Regan's *The Case for Animal Rights*.[2] Regan's arguments challenged my presuppositions.

After rethinking my position, I wrote a trio of essays on ethics and farm animals. Taken together, they tell the story of how I came to give up important background beliefs, and how my new beliefs affected my views about the propriety of making transgenic animals. The first essay, "Pigs and Piety: A Theocentric Perspective on Food Animals," explains why I surrendered deep-seated religious convictions. The second, "The Moral Irrelevance of Autonomy," responds to the most common objection to the idea of animal rights, the objection that while humans are moral agents, animals are not. The third essay, "Should We Genetically Engineer Hogs?" applies the theory of animal rights to transgenic food animals, animals into which humans have inserted foreign genes.

1. Pigs and Piety: A Theocentric Perspective on Food Animals

I live in Story County, Iowa, where the most sustainable way to farm is called family, or mixed, farming. Family farms raise grains in summer and feed them to livestock in winter. Farmers use manure from the animals to supply nitrogen fertilizer to pastures and fields, and they sell the pigs and cows at auction for cash. On mixed farms, the rearing and selling of livestock is the *raison d'être* of the operation, and the operation is, in the current jargon, sustainable, ecologically balanced, and consistent with principles of good stewardship.

Now, I am not a farmer. And I can tell you that Iowans in general are not as virtuous as William Hedgepeth's paean at the beginning of this chapter suggests. But Hedgepeth is right; there are eight times as many pigs in my state as people, and the economic health of grain farms as well as pig farms rests on the practice of raising and slaughtering animals. In the summer, uncle Harold raises corn, soybeans, and hay. All chickens are gone from the farm; the handful of hogs and cattle that remain are little more than Jason's last 4-H project before going off to college. The corn crop is a money maker because it winds up in front of animals at hog confinement, cattle feedlot, or broiler hen operations. Indeed, upwards of eighty percent of my uncle's grain is destined to be fed to pigs, cows, and chickens awaiting slaughter.

Can one question the practice of meat-eating without questioning the institution of the family farm? As I began to reflect on the arguments of the defenders of animal rights, I wondered whether rejecting meat would be equivalent to rejecting the history and identity of the Pippert family. I purposely put the idea out of my mind. It seemed morally insensitive even to envision defending the rights of pigs when economic pressures on farmers were so severe. The playing field was so biased against smaller farmers, and smaller farmers' problems produced so much anxiety, that I found myself wondering what sort of person would ask questions about the well-being of farm hogs when the well-being of farm children was at stake. How could someone who loved family farms reject the central practice on which they are based?

I found, however, that the arguments for vegetarianism are powerful.

1.1 How I Became a Vegetarian

First, I had to decide whether pigs experience pleasure and pain, whether they have emotions, desires, wishes, preferences, a family life. This was not a difficult decision. That pigs are sentient seemed evident to me from watching the pigs on my uncle's farm. Pigs are not, as common knowledge has it, dirty, dumb, or solitary animals. If given a sufficient amount of room, pigs will invariably defecate in the same area, teach their young to keep away from this area, and establish the area at a considerable remove from the sleeping area. Contrary to popular belief, pigs prefer to wallow in clean water, not mud, and will not play with toys soiled by feces.[3] Pigs are intelligent, affectionate, and social animals. The only thing they seem to love more than having their stomachs and ears rubbed is lying next to their neighbors after having run playfully in circles around them, squealing and barking all the while.

What is it like to be a pig? No one can get inside a pig's mind, of course, but we can think carefully about how they appear. Here is William Hedgepeth's perspective on his day spent in a pig pasture:

> Idling hogs amble and squat. Some root. One sneezes. The sleeping hog beside me wags his ear a twitch or two and otherwise remains removed from the milieu. A Hampshire bites a Yorkshire's ear. A Poland China bites my foot. A white hog with a black face and black spot on his side executes a galloping gleeful leap into the vacant pond. A wandering rooter pussyfoots up the hill and sneezes right into the face of the one asleep, who responds merely with another quick ear-wag and continues his snooze (p. 125).
>
> . . . A hog [taking a] siesta on the hilltop has just jumped up to bump an intruding rooter down the slope, somersaulting to the bottom with a tumbling eruption of high-pitched squeals. Most of the hogs are up now, moseying about, perfectly unhurried: gambol and squat awhile, browse in the dried mud, drift in bulky serenity among the stumps and stubble and birds, call a sudden halt to it all every so often to look up at a sound or nudge another in the loin. Probe, poke, trot, root. Ah, hogs! They have unquenchably inquiring minds, each with a vast capacity for sustained wonder (p. 128).

Aristotle believed each animal has a telos or purpose to which it is directed, a "that for the sake of which" it exists. If Hedgepeth is right, the telos of a hog is the will to root, to find his food at least three inches underground, and to get his snout into every tractor tire, hole, and crevice within reach. Not forgetting sleeping and investigating and eating and mating and playing, rooting must be one thing for the sake of which God made hogs.

The daily activities of hogs clearly suggest that they possess desires, preferences, pleasures, pains, and social lives. You may also now have some idea of what the telos of this higher mammal may be. The hog: Kingdom, *Animalia*; Phylum, *Chordata*; Class, *Mammalia*; Order, *Artiodactyla*; Family, *Suidae*; Genus, *Sus*; Species, *Sus scrofa*; Subspecies: *S.s. scrofa* (the Central European wild boar), *S.s. leucomystax* (Japanese wild boar), *S.s. vittatus* (Southeast Asian pig), and *S.s. domestica* (domestic). These are some of the facts about hogs, but facts alone, no matter how many, would never add up to the moral judgment that it is wrong to kill and eat *Sus scrofa domestica*. For that, we need a general moral principle and an argument.

Here is the argument that changed my mind.

a. We may call individuals who are capable of desiring and learning, "individuals with futures."
b. All individuals with futures, such as adolescents, adults, and elderly humans, have a moral right not to be killed for trivial reasons.
c. Most farm animals are capable of desiring and learning;
d. Most farm animals are, therefore, individuals with futures.
e. As humans in developed countries generally do not need to eat meat, meat-eating is a trivial reason to kill an animal;
f. Therefore, combining (b) and (e), animals have a moral right not to be killed for meat by people in developed countries.

When I first started thinking seriously about the one and a quarter inch thick Iowa chops I so loved to barbecue, I thought I had to decide whether pigs had moral rights, and whether I was depriving them of that most basic right, the right to life, by paying other people to carve them up for me. I was impressed by arguments like Joel Feinberg's and Michael Tooley's that it is impossible for an entity to have a right to life unless that entity has interests in the sense of "*able to have* an interest in *x*. "[4] Clearly, it is *in the pig's interest* to be able to sleep, eat, and root. But this is a different, weaker, sense of "interest" than the one required. For there are things that have interests that cannot take an interest in anything. It is in a hay baler's interest to be kept full of baling twine, but the machine does not possess the conscious awareness necessary to take an interest in seeing that it does not run out of twine in the middle of a row. Having things that are in its interest, and even having things that are good for it, does not make a hay baler a bearer of moral rights. The machine does not have the *right* to be well maintained. In order to have moral rights, something must at least potentially be conscious of what is good for it.

The most rigorous philosophical argument for recognizing the moral rights of animals is Regan's *Case*. Regan argues that insofar as humans have basic moral rights because they have desires, are sentient, and have futures, then at least certain animals must have similar rights as well.

Humans have rights because we have intrinsic value; we are subjects of a life, with memories, hopes, goals, social lives, and so on. Insofar as we desire to pursue the interests that make our lives worth living, we have at least a *prima facie* moral right not to be interfered with as we pursue those interests. The most basic moral right is the right of an innocent individual not to be seriously harmed so that others may benefit. Regan, in sum, rejects utilitarian justifications of harm. Now, the fact that many animals have memories, desires, sentience, social lives, and so on, entails for Regan that they are also subjects of a life. Because they are subjects of a life, they have intrinsic value. And because they have intrinsic value, they have moral rights, including the basic one mentioned above. Animals, in short, may not be harmed in order to bring benefits to humans.

Regan puts this argument in different ways, sometimes emphasizing the notion that animals have *interests*. As we have already noted, interests come in two distinct varieties. Welfare-interests are interests of any and every living individual. Plants, for example, have welfare-interests in obtaining sufficient water and sunlight. The second kind of interests, preference-interests, are restricted to conscious individuals. The paradigm of a conscious individual is a normal adult human with the ability to form and modify desires. Now, individuals with welfare-interests do not necessarily have moral rights. If they did, then we would have to grant moral rights to plants, bacteria, and the strep virus. That is counterintuitive. Every human, on the other hand--or at least every human with the ability to form and modify desires--has preference-interests. These interests are critical to our identity and, assuming that our desires are peaceful and involve no harm, we have at least a *prima facie* moral right to pursue them. When Regan formulates the animal rights theory in terms of interests, he is building on the idea that, all other things being equal, an individual with a preference-interest in doing *x* has at least a *prima facie* right to do *x*, so long as doing *x* harms no one.

Adult mammals have preference-interests and many of these preference-interests are non-maleficent (that is, their satisfaction involves doing no harm to any sentient being). Insofar as animals have basic peaceful desires--to acquire food, water, shelter, companionship--then, according to the principle of fairness, these animals must have the analogous *prima facie* basic moral right that any human has in the human pursuit of these goods. Insofar as we recognize moral rights for humans on the basis of our being subjects of a life, having intrinsic value, or possessing preference-interests, we must also recognize moral rights for every individual who is the subject of a life, has intrinsic value, or possesses preference-interests.

Regan believes that the consequences of the animal rights theory (AR) are radical, and he demands the end of the practice of raising and slaughtering of animals for meat, the end of hunting, of rodeos, and of zoos. Writing that "you don't change unjust institutions by tidying them up," he extends the implications of the theory to "the total abolition of the use of animals in science. "[5]

To my mind, the strongest argument against Regan's position is that of R. G. Frey. Frey claims that animals cannot have moral rights because rights require interests, interests require language, and language requires concepts. Frey denies that animals have concepts; therefore, animals cannot have language, interests, or moral rights. According to Frey, an individual must possess *concepts* in order to possess interests in the relevant sense because if one lacks concepts, one cannot represent anything to oneself. And if we cannnot represent anything to ourselves, how could we possibly take an interest in anything? Without concepts and language in which to formulate them, we would have no conceptual tools by which to formulate, much less pursue, interests.

Do pigs lack desires, concepts and language? When my uncle's barrows and gilts lift the lids on their feeder bins, there is hardly a simpler or more efficient way of interpreting their behavior than to say that they *desire* to eat. When Hedgepeth's piglets chase each other around the pasture, there is no better explanation than, "the pigs *want* to play." The conceptual scheme of beliefs and desires is as apt an explanatory scheme for animal behavior as for human behavior.

Assuming that pigs have beliefs and desires, the next question is whether they have concepts. If they *believe* that there is food under the lid, or that by hiding behind the tire they will surprise their buddies, then it would seem that they must possess concepts, because beliefs are made of concepts such as "food" and "over there." If animals have concepts, then they may be capable of *taking an interest* in their activities. And if they can take an interest, they are at least potential bearers of something like a *prima facie* moral right to pursue their interests, assuming that they are not harming any other beings by so doing.

Frey is convinced that animals do not have concepts or language because they are not capable of making assertions or lying. It follows, according to Frey, that painless slaughter does not violate a pig's right to continued existence because pigs, lacking concepts, cannot have language; lacking language, cannot have interests; and, lacking interests, cannot have moral rights.

This line of argument, if sound, would constitute a powerful philosophical justification for the historical practice of domesticating and eating pigs, and would buttress agrarian positions that emphasize ecological harmony and stewardship of nature. But there are two questions here: Must a being have language to have concepts? And, Do animals indeed lack language?

Consider the second question. As far we know, pigs do not have "the ability to make or entertain declarative sentences," Frey's way of interpreting what it means to have language. But pigs communicate with each other, and they can communicate with us in limited, distinctive, ways.[6] Pigs, moreover, appear to many observers to reflect in a self-conscious way about their environment. Some, including me, think they have seen pigs trying to deceive each other. Here is a reason for thinking that some vertebrates, at a minimum, have language.

But suppose that we are wrong in this judgment. Suppose that pigs lack language; even so, they may still possess concepts. Here is an argument, suggested to me by my colleague Bill Robinson, showing that individuals use concepts even in the absence of language.

To use a concept is to classify something, to represent a difference to yourself. One sign of the ability to represent a difference to yourself is the ability to exhibit reliable differences in your disposition to behave, reliable differences in behavior that are correlated with (other) differences in your environment. There are empirical ways to determine whether individuals without language are nonetheless capable of exhibiting reliable differences in their disposition to behave correlated with differences in the environment. To determine whether individuals lacking language are capable of exhibiting reliable differences in their disposition to behave correlated with differences in the environment we must observe their behavior and its correlation to the environment. Then we must determine whether differential dispositions to behave are correlated with differences in the individual's environment.

Start with the human case. Assume that an otherwise typical adult, Jim, cannot talk about his behavior, not even to himself. Nonetheless, we notice that Jim sometimes waves his arms and points to his mouth and sometimes he does not. Upon examination, it becomes clear that when Jim is waving his arms and pointing to his mouth, he has not eaten for two hours or more. Anytime that he has eaten within two hours, he does not point to his mouth.

We can correlate Jim's differential disposition to behave (waving or not waving) with a difference in his environment (having eaten or not having eaten). Differential behaviors we observe must be reliable differences as ascertained in a proper scientific way, for example, observed under the strict conditions of double blind observer experiments. That is, there must be no problem in telling that there *is* a difference in the individual's behaviors. Lacking language, Jim nonetheless uses concepts to represent his hunger to himself and others.

We would not deny Jim basic moral rights, such as the right not to be deprived of food when he is hungry and there is plenty of food to give him. Why should we not apply this same analysis to animals that may lack language? There is ample empirical evidence that animals exhibit reliable differences in their dispositions to behave, and that these differential dispositions to behave are correlated with differences in their environment. Therefore, animals are capable of exhibiting reliable differences in their disposition to behave that are correlated with differences in the environment. It follows that animals are capable of representing differences to themselves. And it follows that animals are capable of using concepts, even if they lack language.

The upshot of this argument is that even if we grant Frey's claim that animals lack language, it is still the case that they have the capacity to use concepts. And, as concepts are constituents of beliefs, and beliefs are constituents of moral rights, then animals have the capacities needed to be bearers of moral rights.

If my claims about pigs' mental states are correct; and if the moral principle that it is wrong to deprive a being of its right to life is defensible; and if I have made no mistakes in reasoning to the conclusion, then it may be wrong to deprive a pig of its right to life for a trivial reason. But how does respect for individual animals fit with an overall theory of environmental ethics?

1.2 Animal Rights or Environmental Ethics?

Paul Taylor makes the attitude of respect for nature the basis of all moral reflection about the environment, and identifies four dimensions of that attitude. Two of them are relevant here. The first is the valuational dimension, "the disposition to regard all wild living things in the Earth's natural ecosystems as possessing inherent worth."[7] The second is the affective dimension, "the disposition . . . *to feel pleased about* any occurrence that is expected to maintain in existence the Earth's wild communities of life, their constituent species-populations, or their individual members."[8]

Taylor believes we owe the attitude of respect toward wild living things. He avoids the language of animal rights, but he insists we follow the principles of proportionality and minimum wrong.

The first principle means that we should never act disproportionately, for example, violating an elephant's basic interest in life simply to satisfy our nonbasic interest in having ivory carvings on our mantlepiece. "Greater weight is to be given to basic than to nonbasic interests, no matter what species, human or other, the competing claims arise from. Nonbasic interests are prohibited from overriding basic interests."[9]

The second principle states that "the actions of humans must be such that no alternative ways of achieving their ends would produce fewer wrongs to wild living things."[10] From these two principles you may see how protective Taylor is of wildlife. His attitude toward domestic animals is less than clear, however. The reason is that Taylor is impressed by the fact that pets and food animals have been purposefully bred to serve a human purpose.

Unlike wild animals whose existence does not depend on their fulfilling our needs, domestic animals exist only because we have exercised dominance over them and their environment.

Taylor puts the matter forcefully. The practice of rearing food animals depends, first, he writes:

> on total human dominance over nonhuman living things and their environment. Second, [it involves] treating nonhuman living things as means to human ends . . . The social institutions and practices of the bioculture are, first and foremost, exercises of absolute, unconditioned power . . . When we humans create the bioculture and engage in its practices we enter upon a special relationship with animals and plants. We hold them completely within our power. They must serve us or be destroyed. For some practices their being killed by us is the very thing necessary to further our ends. Instances are slaughtering animals for food, cutting timber for lumber, and causing laboratory animals to die by giving them lethal dosages of toxic chemicals.[11]

While Taylor does not explicitly draw the conclusion that it is morally permissible to continue to subdue nature in this way, this conclusion is implied in his remarks. Other environmental philosophers, including Callicott and Midgley, have a similarly bifurcated attitude toward animals.[12] Wild animals should, other things being equal, be allowed to live unless they are being hunted for food. Domestic food animals, on the other hand, are intended for slaughter.

1.3 Have We Created the Domestic Hog?

It began to look as if my evaluation of Taylor's environmental philosophy might cause me to overturn my decision against meat eating. If there is an absolute difference between wild and domestic animals, and if this difference means that wild animals have intrinsic value while domestic animals have only instrumental value for humans, then it might be permissible to raise and slay hogs and yet impermissible to kill wild wart hogs. To decide whether the difference between tamed and untamed was really this decisive, I had to read some animal science. Just how different are Minnesota Number Threes from wild boars?

I immediately ran into a problem. To my knowledge, there are no scientific studies comparing the physical or behavioral traits of specific domestic pigs with wild pigs. Nonetheless, on the basis of certain generalizations scientists have proffered in the literature on swine production, some observations about the difference can be offered tentatively.[13]

Wild pigs tend to have aggressive dispositions. They often live in herds of four to twenty foraging animals consisting of one or more adult females and their young. Boars range freely in forest settings throughout the year, staying close to the herd during the reproductive season, when they become territorial and protective. Omnivorous and voracious eaters, sows and boars alike spend the majority of their waking hours walking, rooting, and eating. The courtship of an oestrus female by a wild boar lasts several days, with the male grunting a soft rhythmic mating song and having to overcome a last minute rebuttal from her when she wheels and faces him just before he tries to mount her. The wild sow may spend days making a nest for her young. The boar seems to enjoy the presence of piglets, tolerating them as they wiggle on top of him as he rests.

Domestic swine tend to be larger, less fatty, more docile toward humans and less agonistic toward each other. As you might guess, we have little information about how large a "domestic herd" might be because pigs in confinement are not allowed to form natural social groups. Boars are kept away from the sows, feeder pigs are thrown together according to age, and sows are kept in maternity pens before parturition and during nursing. Even though they are usually denied the space and freedom to form natural relations with other pigs, domestic pigs are still known to adapt rapidly to new conditions. They exhibit a high degree of intelligence and have, for example, been trained to hunt truffles and indicate targets like Pointer dogs.

The sexual relationships of confined pigs are noticeably different from their wild counterparts. When a sow in heat is presented to a boar, copulation occurs quickly. There is very little behavior corresponding to the long courtship of wild sows and boars, as domestic sows usually allow boars to mount immediately, and boars are selected, in part, for their virility and promiscuity. Boars kept away from sows sometimes form stable homosexual relationships. Their behavior toward young piglets is hard to observe for reasons noted above.

There are, in sum, significant differences between the physical, psychological, and social characteristics of domestic and wild pigs. Wild pigs tend to be smaller, fattier, more romantic, less promiscuous, and more ferocious. Domestic pigs tend to be larger, leaner, less romantic, eager to mate in season or out of season, and more docile. The differences stem from the influence of human intervention as farmers have consciously selected individual pigs for the traits now possessed by sows and boars. Breeders have weakened the pig's natural defenses, and rendered them dumber, less agile, and more meaty, than their wild relatives. Differences are undeniable. And yet we may ask, how great are the similarities? Are the differences significant enough to justify claims that we have exercised "absolute power" over the domestic animals?

The differences in physical appearance of African bush pigs and Duroc hogs are noticeable, but both look more like the other than they look like other species. Both adapt quickly to changed environmental conditions. Both exhibit tremendous behavioral plasticity in the face of fluctuations in weather, diet, and physical threats. Both exhibit attitudes of defiance, pride, and affection. Both are extremely social. Both prefer not to leave the company of others, except for the case of older males, who sometimes prefer occasional solitude. Both like to root in soil and water, to wallow in pools. Both exhibit distinctive territorial behavior, keep separate areas for elimination of urine and feces, and train their young to do the same. Both are curious about new objects, and will sniff and nibble any protrusion or hole. Both have a complex range of vocal snorts and whoofs for communicating a variety of emotions, signals, and alarms. Both have nearly identical olfactory and auditory capacities. Neither is able to regulate body temperature for at least two days after birth. Neither is receptive to newcomers to the herd. Both are gregarious animals, huddling together against cold weather and enjoying warm weather in close proximity.

The list could go on, but the point has been made; the differences between domestic and wild pigs pale in light of their similarities. May we then continue to believe that we have exercised "unconditioned" power over the being of the production hog? The scientific evidence fails to support the claim because the identity of the production hog is as much a product of natural forces as it is of human intervention.

May we at least claim responsibility for the distinctive features for which we have selected in our hogs? For example, domestic pigs are diurnal creatures whereas wild pigs sleep during the day and are active at night. Is this trait a human mark stamped on the pig? It may be, just as the sexual promiscuity, docility, and physical size of the domestic hog may be marks of human intervention. Still, we must ask whether these traits are really of our doing or whether they are not responses that may be equally attributed to the hog. Consider that domestic hogs tend to be diurnal creatures whereas wild hogs tend to be nocturnal (hunting is easier in the evening hours.) Did humans cause this difference? I doubt it. Hogs are highly adaptable creatures, and there is not much stimulation in hog pens at night. The domestic hog's preference for daylight activity may be a tribute to their own plasticity of behavior, a trait caused as much by the pig's own initiative as by the breeder's selections. Being diurnal, in short, may be a learned response to environmental conditions, and it may be a characteristic pigs would abandon if turned out of their pens or if stimulated at night. This suggests that certain behavioral differences between domestic and wild species may not only not be permanent but may be reversible.[14]

Based on a review of the empirical differences between undomesticated and domesticated hogs, Taylor's claim that we have created these animals seems weak, as does the implication that they are human artifacts we may regard as our tools.[15] Today's breeds are expressions of human power and control over nature, the result of invasive, repeated, and sustained manipulations of generations of animals. The Durocs and Hampshires and Yorkshires now on Mennonite family farms would almost certainly not be here were it not for humans. Hogs are part of our moral community in a way wild animals never have been because their evolution is intricately connected with our own. They depend on us for their existence. But it does not follow that we are justified in continuing to intervene in their histories by encouraging them to inbreed, and by slaughtering their young.

If Taylor's views about food animals are not entirely clear, other environmental philosophers' views are clear. Midgley and Callicott seem to condone meat eating as part of the long history of relations between humans and domesticated animals. The view gains credence in light of the fact that the history of a being is relevant to deciding what that being is and what our natural duties are toward it. Consider Midgley's view. She approaches ethics from a biosocial perspective, and points out that we are members of nested communities, each of which has a different structure. According to our various roles in the various communities, we have various duties. The central community for many of us is an immediate family. We have duties not only to feed, clothe, and shelter our children, but to bestow affection on them. Bestowing similar affection on our neighbors' children is not similarly required of us, however. Not only is it not our duty, but, as Callicott observes, "it would be considered anything from odd to criminal" were we to behave toward neighborhood children the way we behave toward our own.[16] At the next level, we have "obligations to [our] neighbors which [we] do not have to [our] less proximate fellow citizens--to watch their houses while they are on vacation, for example, or to go to the grocery for them when they are sick or disabled." And then we have obligations to those in our state "which we do not have toward human beings in general, *and* we have obligations to human beings in general which we do not have toward animals in general."[17]

These subtly shaded social-moral relationships are complex and overlapping. Thinking of animals, Midgley argues that pets are surrogate family members and merit treatment not owed either to less intimately related animals, for example, to barnyard animals, or to less intimately related human beings. Following Midgley's biosocial line of thinking, the narrative history of each animal defines its identity. Since hogs have been bred to play a certain role in our community, our duties toward them derive from understanding what their role naturally is.

Like Midgley, Callicott argues that the welfare ethic of the mixed community enjoins us to leave wild or "willed" animals alone, while caring humanely for domestic species. This means that we are justified in using domestic animals in the ways they have been bred to be used. It is not inhumane to use a Belgian draft horse to pull a wagon, as long as you do not abuse her in the process. It is not inhumane to kill pigs and chickens and steers for food as long as you care for them in a way that does not violate the unspoken social contract we have evolved between human and beast.[18]

Reading environmental philosophy made me wonder whether my decision not to eat meat had been divorced from narratives, history, and common sense, in the worst way. If the history and social role of a being plays a decisive role in determining what that thing is, and if today's pigs would not be here if it were not for the long history of human intervention in the mating patterns of hogs, then the raising and slaughtering of pigs is the very practice necessary for Durocs, Hampshires, and Minnesota Number Twos to exist at all. Who was I to condemn these creatures?

Callicott seemed to press the point on me. Those who condemn meat eating thereby condemn the "very being" of the animals they are trying to defend. For without the long historical practice of meat eating, Callicott writes, these particular animals would not exist.

But must we condemn someone's existence if we disapprove of the lifestyle they are forced by others to lead? To condemn the way someone is treated is not to condemn them. We condemn a life of forced prostitution without thereby condemning the prostitute. In the interest of the good of the prostitute, we condemn the power relationship that has come to restrict her freedom. Analogously, we can condemn the practice of domesticating and slaughtering pigs without thereby devaluing the existence of *the* pigs.

Having answered Callicott's challenge, I went back to Taylor's rigid differentiation between the respect owed wild animals and his quasi-instrumentalist view of domestic animals. I discovered on second reading that, despite his dismissive attitute to food animals, he insists nonetheless on vegetarianism.[19] His reason has nothing to do with the individual animal's worth or rights, however. It is based instead on an environmental principle of fairness, captured in the metaphor of sharing the earth. Taylor draws attention to

> the amount of arable land needed for raising grain and other plants as food for those animals that are in turn to be eaten by humans when compared with the amount of land needed for raising grain and other plants for direct human consumption . . .

> In order to produce one pound of protein for human consumption, a steer must be fed 21 pounds of protein . . . [a pig must be fed] 8.3 pounds . . . [and a chicken] 5.5 pounds.[20]

Taylor would have us return the land now in cultivation to grow grains for cows and pigs to wildlife refuge.

Taylor argues for vegetarianism on ecological grounds, lamenting the fact that humans have taken over much more than their fair share of the temperate regions of the globe. To return land to wild animals we should cultivate less ground, shrink our farms' size, and concentrate them in local regions so as to leave larger tracts of wilderness. Thus, even though they are not grounded in the theory of animal rights, environmentalist reasons are offered by Taylor for abolishing the practice of meat eating.

1.4 Animal Rights as an Environmental Ethic

I still had two questions: Would it be wrong, if we pulled in our plows and chemical sprays and shared the earth equitably with other species, to eat an occasional future pig raised on a small nonfactory farm? And if in that ideal world some of us revert to hunting and gathering as a permanent lifestyle, would it be wrong for us to kill and eat one of the millions of wild pigs?

To answer this question, I went back to Taylor's five priority principles. When the requirements of human ethics compete with those of environmental ethics, Taylor tells us to follow principles exhibiting the attitude of respect for nature. The fundamental criterion is fairness, read as species-impartiality. According to Taylor, both plants and animals deserve respect, even though neither one is a primary moral rights holder. The first priority principle is the principle of self-defense.

> *It is permissible for moral agents to protect themselves against dangerous or harmful organisms by destroying them (p. 264-265).*

This principle "condones killing the attacker only if that is only way to protect the self." We must "choose means that will do the least possible harm (pp. 265).

The second principle is the principle of proportionality, and it deals with conflicts "between *basic* interests [for example, food, water, and continued existence] of animals/plants and *nonbasic* interests [for example, air conditioned offices] of humans."

> *Greater weight is to be given to basic than to nonbasic interests, no matter what species, human or other, the competing claims arise from. Nonbasic interests are prohibited from overriding*

> *basic interests (p. 278).*

This principle prohibits such practices as

- Slaughtering elephants so the ivory of their tusks can be used to carve items for the tourist trade.
- Killing rhinoceros so that their horns can be used as dagger handles.
- Hunting and killing rare wild mammals, such as leopards and jaguars, for the luxury fur trade.
- All sport hunting and recreational fishing (p. 274).

The third principle is the principle of minimum wrong. Like the second principle, it concerns conflicts "between basic interests of animals/plants and nonbasic interests of humans."

> *The actions of humans must be such that no alternative ways of achieving their ends would produce fewer wrongs to wild living things (p. 283).*

Plants and animals and humans have equal inherent worth, in Taylor's estimation, but he recognizes that rational people may decide to engage in activities involving harm to wild living things. As long as these people are "rational, informed, and autonomous persons *who have adopted the attitude of respect for nature*," then "it is permissible for them to pursue [their] values only so long as doing so involves fewer wrongs (violations of duties) than any alternative way of pursuing those values" (pp. 282-283).

Taylor's fourth principle is the principle of distributive justice, and applies to "conflicts between *basic* interests, in which nonhumans are not harming us." The cases in question, then, are cases where the principles of self-defense, proportionality, and minimum wrong do not apply.

> *When the interests of parties are all basic ones and there exists a natural source of good that can be used for the benefit of any of the parties, each party must be allotted an equal, or fair, share (p. 292).*

The fifth principle is the principle of restitutive justice:

> *When harm is done to humans, animals, or plants that are harmless, some form of reparation or compensation is called for. The greater the harm done, the greater the reparation required (p. 304).*

Using these principles, I was able to answer my two questions.

Consider the second question first. If I lived in a place or a time where I could not survive without hunting wild goats and sheep, or fishing for tuna and whales, then it would be permissible for me to kill and eat those animals. Why? Because the first principle enjoins self-defense and, *per hypothesis*, the only way to protect myself from death under the circumstances would be to hunt or fish. As long as I kill in a way that respects the principles of fairness, minimum wrong, and proportionality, I will be justified in my carnivorous behavior. There is, Taylor sagely points out, no principle requiring me to sacrifice my life for the sake of animals.

Consider now the first question, whether raising and slaughtering animals would not be permissible in the ideal world, in the world where the number of humans and farms is dramatically reduced. If there were, say, only 500 million of us instead of 5 billion, and only 50,000 small farms instead of half a million corporate farms, then other species might flourish. Under those conditions, couldn't rational autonomous persons who have adopted the principle of respect for nature decide to raise pigs in such a way that the animals were allowed maximal freedoms and long unhurried lives? And wouldn't it then be the case that those animals would be better off living *that* lifestyle than never having the opportunity to be born at all?

This question is more difficult, but it seems to me that we should answer it negatively. The principle of self defense could not be enjoined to sanction such activity, because slaughtering the pigs in question, even toward the end of their lives, would not serve any *basic* interest of ours; we can get our protein elsewhere. The principle of proportionality also offers little support, because our nonbasic interest in enjoying a good set of barbecued back ribs is prohibited from overriding the pig's basic interest in continued existence. The principle of minimum wrong would also argue against even a low level of meat eating, since there are alternative ways of achieving our interest in experiencing robust gustatory pleasures.

Careful consideration of the natural relations of all things and rigorous adoption of the attitude of respect for nature inclines strongly toward moral vegetarianism. And thus was I moved, against my personal convictions about the virtues of family farms, to think some higher mammals have mental lives roughly analogous to ours; that killing them for food, even in a painless fashion, does harm to them; and that I should stop having bacon for breakfast.

1.5 Meat-eating as a mutual covenant

I want to say something against three arguments for meat eating. The first two arguments can be dealt with briefly, but the third will call for extended discussion.

The first argument has been admirably formulated by Midgley and Callicott, among others, who claim that the domestication of animals is a mutual covenant evolved between animals and us. Our obligations to animals are therefore determined by our evolved relation with them. The idea here is that animals do not simply serve us; we have a contract to provide them with food, water, shelter, care, and comfortable lives. Some animals, such as our pets, are close to us, and we owe them more than we owe more distant animals.

But what is the responsibility of so-called food animals in this nested hierarchy of evolved relationships? To pay us back with their lives at an early age simply to satisfy our pleasure in eating their carcasses. The contract seems a bit one-sided. The argument would make more sense if it was generally understood to mean "Let the animals live in their natural social groupings, provide them with conditions under which they can pursue their interests, and let them live until a ripe old age before slaughtering them."

But that is not the way the alleged covenant is generally understood. We squeeze hogs together into pens not large enough for them to establish their own area for defecating, we throw them together into new social groupings every few weeks, we control their reproductive cycles with manufactured drugs, and we kill them before they are six months old. If the terms of the agreement were to support hogs into comfortable retirement and then take the carcasses of animals dying of natural causes for sausage, the covenant argument would be more persuasive.

1.6 Killing as a spiritual practice

The second argument is that killing animals is permissible as long as we do it in the right way. Native Americans kill the buffalo with a tragic sense for the loss of its life, and they kill only the number they need. They either eat or use the entire animal, and they do all of this with a humble and grateful spirit, demonstrating respect for the harmony and balances of nature.

Is it permissible to kill and eat animals this way? Here Taylor's response seems appropriate. If it is a question of survival, if it comes down to my life or the buffalo, then the principle of self-defense justifies the killing. However, few people reading this book face such dire circumstances.

The third argument is that humans are morally superior to animals because we have a key characteristic animals lack: free will. This objection is powerful.

2. The Moral Irrelevance of Autonomy

Frey has argued that the possession of "moral rights" is not the line separating us from nonhuman animals. His reason is not that some animals are inside this line (he denies that any are), nor that some humans are outside it (he affirms that many are), but rather that the line itself is too fuzzy.[21] Talk about moral rights, Frey explains, is unsupported by good arguments and is more successful as rhetoric than as philosophy.[22] Frey is a utilitarian who puts little stock in general in the Kantian picture of morality. Consequently, he rejects the idea that any beings have moral rights.

In an article titled "Autonomy and the Value of Animal Life," however, he argues that nonhumans lack moral standing not because they lack what no one possesses (moral rights).[23] Rather, animals lack moral standing because they lack what all "normal adult humans" possess: autonomy, the ability to control or make something out of our lives.[24]

Why should Frey want to shift the burden of the case against animals onto the back of a concept traditionally associated in the most intimate way with that of moral rights? Because he finds it a far less ambiguous notion, not to mention a less controversial one. In the first sentence of the article he claims that autonomy has had "great stress" placed upon it: "in Anglo-American society, [by] virtually every moral theory of any note."[25] Because it has received such stress by so many other theorists dealing with so many other kinds of ethical issues, he believes that it may serve as the limiting concept for all inquiries into our moral duties toward animals.

Frey is not alone in focusing attention on this line; Regan's case for animal rights puts as much weight on autonomy as Frey's case against animal rights. Regan's strategy is to try to show that adult higher mammals are autonomous in an important sense. Thus, Regan makes each of the following claims: that many animals "have preferences and have the ability to initiate action with a view to satisfying them;" that this constitutes "preference autonomy;" and that many animals, possessing such autonomy, must therefore be granted moral considerability.[26] Regan does not agree with Frey as to where the line should be drawn, but he does agree that autonomy should play a crucial role in determining moral standing.

Frey is convinced that "the way is . . . open" to killing and eating beings that are not autonomous.[27] Is he right?

Frey's definition of autonomy is narrower than Regan's understanding of autonomy as "preference autonomy." It has three elements, the first of which is the freedom to act on our own behalf. Autonomy is "our desire to achieve things for ourselves," to make "something of *our* lives," the way a fledgling philosopher might want to succeed on her own rather than trying to ride on her famous husband's coattails. To illustrate the point Frey tells of an academic acquaintance who was concerned that his untenured wife might not be promoted. The husband suggested that he write some publishable papers which she could take and revise and then submit to journals as her own. The woman was rightly insulted by the idea because she did not want to make something of herself by deceitfully using her husband's work. She wanted to make something of herself by relying on her own talents and powers. *She* wanted to make something of *her*self. By rebuffing her husband's attempt to intrude, the woman showed that she was not subject to control by paternalistic outside forces. She was free "of the coercive interference of others. "[28]

The second requirement is freedom from internal coercion. In order to pursue the ends we most cherish we must not only gain independence from the desires of outsiders but we must master our own desires as well. "A certain ordering" of life is necessary if an untenured professor is to "put herself in a position to be able to produce serious academic work."[29] If she does not control her minor impulses she will be pulled in so many directions that she will not be able to devote herself to the desire she desires most. Self-government means that we are able to forego certain lower-order preferences (e.g., playing in a city basketball league) in order to pursue higher-order desires (e.g., making associate professor). Freyan autonomy requires "internal" as well as "external" freedom, the ability to make higher-order decisions about the relative importance of lower-order desires.

The third requirement is to decide for oneself about the kind of life one wants to lead. The professor who successfully resists the intrusions of her husband and who successfully controls her less desirable desires may still be doing something she has not chosen. Suppose that she is working to be associate professor for no other reason than that her mother was a professor before her and her grandmother before that and she feels, for religious reasons drummed into her as a child, that she ought to do what her family wants. Frey would not call this woman autonomous because she is not pursuing a career she has chosen for herself. She is pursuing a plan of life that has been imposed upon her. Notice that she has all of the equipment needed to survey a range of possible plans and to select one for herself but simply has not used it. Instead, she has settled for doing the best she can in what she considers "the family's" line of work. Freyan autonomy requires that we think rationally about the variety of conceptions of the good life, deliberately choose one, and consistently pursue it.

Being in control is central to Frey's theory, as his example of a nonautonomous person shows. Imagine a successful businessman who longs to be a painter and yet continues to spend his energies perfecting his father's business. Frey's opinion of such a man is harsh, and he thinks many of us will "doubtless" be struck by how "weak" the man is. Frey puts the matter straightforwardly: "the real charge against this man is servility; he has allowed, for whatever reason, others to impose their conception of the good life upon him."[30] Here we see how much weight Frey attaches to the third requirement. You are not autonomous: if you have not selected a plan of life from a range of options; if you have not made up your mind about what you think the good life is; and, if you have not taken decisive action to pursue your conception of the good life.

Those who are not autonomous, Frey believes, are morally inferior to those that are autonomous. Denying that all humans have equal moral value, he asserts that the value of someone's life is directly related to its quality.[31] Since he thinks that the quality of the moral life of a nonautonomous person is less than the quality of an autonomous person, Frey must also think that we would all be better persons morally speaking if we seized control of our lives, took matters into our own hands, and changed careers to pursue the one we most desire.

Frey does not address himself to some of the knottier questions raised by his analysis. Is autonomy intrinsically good or good as a means to another end? Frey seems to think that it is good in itself. But can't we develop our autonomy at the expense of others? Couldn't we strive to become more autonomous in order better to exploit others sexually or coerce them into unearned business favors? Nor does Frey tell us what to think about moral theories in which autonomy has not been heavily accented. Such theories may not be part of something called "the" tradition of Anglo-American moral theory but they are undeniably part of the moral practices of many followers of the Land Ethic and other forms of moral environmentalism; Natural Law, Divine Command, and other religiously based theories; feminist and pragmatist perspectives; and aretaic theories.

Nor does Frey address in this article the most troubling question of all: Even if autonomy were demonstrated to be the line separating us from animals, would that justify cementing baboon heads into steel sleds and slamming them against walls? May we so treat any and every being that lacks autonomy?

However urgent these questions may be, they are not ones Frey sets out to answer in the present essay, and I will not pursue them here. Rather, the central claim of his paper is that autonomy is a property of the "normal adult human" and a necessary feature of the good life. It is this claim I wish to contest. Frey could mean it in one of two ways. He could intend it as a descriptive claim, that all "normal adult humans" just are autonomous. This would be an empirical judgment about the kinds of lives led by most people in the world. If this were Frey's intent we would have to do some social-scientific work to find out whether he was right. Lacking the results of such a study and basing my response only on my own experiences with what appear to me to be "normal adult humans," I must nevertheless say that I find this view fantastic. The majority of "normal adult humans" I know are far from autonomous in Frey's sense, and shortly I will try to describe a nonautonomous, morally valuable, person.[32]

But Frey might intend his claim, on the other hand, as a normative judgment--that all normal adults *should* be autonomous. This is more properly a philosophical judgment, and one with which I disagree. I do not believe that autonomous people necessarily live lives of higher moral quality than less autonomous folk and the person I will describe below will serve to show why I hold this view as well.

Assume that autonomy is, on the whole and all things considered, a good of one sort or another. In the absence of other considerations, it is better to have control over your life than not to do; better to have a life plan than not to have one; better to be internally free than to be tied up by your lesser desires; better to be externally free than to be hamstrung by others' plans for you.[33] Assume further that "the value of life is a function of its quality, its quality a function of its richness, and its richness a function of its scope or potentiality for enrichment." And assume too "that many humans lead lives of a very much lower quality than ordinary normal lives, lives which lack enrichment and where the potentialities for enrichment are severely truncated or absent."[34] From these premises it does not follow, as Frey seems to assume, that beings who are not fully autonomous are beings who either lack moral standing altogether or who would have a higher quality of life if they exercised more control over it.

To see the fallacy of the conclusion consider a normal adult human who lives a life of high moral quality but has never formulated a plan of life. George is a fifty-seven year old father of six who not only can "read, do higher mathematics, build a bookcase, [and] make *baba ghanoush*," but who has driven a truck across country for twenty five years.[35] Graced with superior counselling skills, George is an excellent conflict mediator, known for quietly but effectively intervening between tired colleagues in diners from coast to coast. He is adept at smoothing out the edges of a difficult way of life in ways that are no less significant for being nigh imperceptible.

George likes his job and is good at it. But he did not choose his career. While he finds some measure of fulfillment in being a driver, he would rather play golf semi-professionally and volunteer his time delivering meals to elderly folk around town. He hesitates to quit his job, however, because he fears losing seniority, a very good income, a measure of self-fulfillment, and reasonably happy working conditions. Moreover, he does not really know how he would go about "changing careers" at this point, and he believes (almost certainly in error) that his wife and teen-aged children are not in a position to afford him that luxury. Above all else, George wants his children to be happy and his wife satisfied. His perception of their needs is more important to him than his other career desires.

Being a father is the activity that gives George the most satisfaction. But is this a deliberately chosen higher-order preference? Surely it is for some men, but this does not seem to be the right way to describe George. George is a reflective and skilled person who has shaped the lives of others in profound ways. But, as he says himself, his satisfaction in parenting is more instinctual than chosen. Raised in a rural area by conservative Catholic parents, George's conception of the good life is more an inherited one than one he has deliberatively chosen from a menu. He never remembers having thought about, much less deliberately chosen, a "plan" of life.

Not only does George fail Frey's three-fold criteria for being in the autonomy circle, but he does not want to try to get in. George has paid careful attention to the cultural conditions in which his children were raised and he is not at all certain that he approves. He has known for a long time how strongly they were encouraged from kindergarten on up to "find themselves," to exhibit independence of thought, to formulate a rational life plan, to seek equality with others, to pursue their own happiness. Sometimes he finds this amusing because when he was growing up "you didn't have all this agonizing over who you were and where your 'relationships' were going--you just found a woman, fell in love, and got married." But other times he is profoundly disturbed by it. He fears that his children have been coerced by their consumeristic culture into placing an overweening importance on their own successes, their own achievements. Being happy is their bottom line. When George was growing up, that was not the bottom line; it was caring for others. By allowing his children to chase autonomy has he also let them lose sight of the value most cherished by *his* father?

George is a full moral agent with immense talents in the areas of care, compassion, hospitality, fairness, discernment, responsibility, loyalty, and love. He exhibits, in short, an extremely high quality of moral life. Notice that he is by no means a "less" rather than "more" normal human, much less a marginal one. George is as normal a human as you can find. If he has any distinction, it is only that he is such a good person. Nevertheless, he has not selected a plan of life from a range of options; he has not made up his own mind about what the good life is; and he has not taken decisive action to pursue his conception of the good life.

George is a reflective, nonautonomous, saint. He does not have control over his life and, moreover, he is incapable of exercising control over his life. And yet the quality of his life is extremely high. Here is a man who falls outside of Frey's circle.

To get a clearer picture of the type of individual I have in mind, consider four different types of saints.[36] A saint is anyone who lives a self-sacrificial life. A self-sacrificial life is one that consistently promotes the legitimate interests of others while, on occasion, acting contrary to legitimate interests of the self. *Strongly autonomous* saints act self-sacrificially because they want to act self-sacrificially. Mother Teresa of Calcutta not only has the ability to reflect critically on her desires but the freedom of will to change her way of life if she decides, one morning, to leave the poor and get into advertising. *Weakly autonomous* saints act self-sacrificially because they want to act self-sacrificially. Unlike the strong autonomous saint who energetically affirms and reaffirms her way of life, however, this saint is attracted to other conceptions of the good life. She is unable to switch directions because she lacks the willpower to act on her other desires. She continues to act self-sacrificially, but this is as much because of weakness of will as anything else. Both of these saints fit Frey's depiction of what he calls normal adult humans. Both are autonomous. And the weakly autonomous saint would be better off if she were to take more control over her life.

Now consider two nonautonomous saints. The *strongly nonautonomous* saint acts wantonly in a self-sacrificial way. Like Felicité in Flaubert's short story, "Un Simple Coêur," this saint's operative desire is always to relieve the suffering of others, but the desire to relieve suffering is not a desire she has chosen. What is more, this is not a desire she could choose, because nature and nurture have conspired against her to produce a person who lacks freedom of will. In Harry Frankfurt's apt expression, she "neither has the [will she] wants nor has a will that differs from the will [she] wants. "[37] Driven by psychological and sociological forces beyond her control, Felicité just happens to be a saint rather than a sinner.

Weakly nonautonomous saints, like strongly nonautonomous saints, do not have the power to choose their self-sacrificial way of life. Their operative desires are out of their control, determined by powerful forces of behaviorial make-up, habit, and socialization. Unlike the strongly nonautonomous saints, however, these persons are conscious of the forces shaping their lives and are capable of reflecting on their desires. They are sometimes disposed, like George, to want a different way of life. Unfortunately, they lack the willpower to act on these desires. Like the weakly autonomous saint, the weakly nonautonomous saint is not always happy with the fact that he is a truck-driving saint instead of a golf pro.

George is a reflective, weakly nonautonomous, saint. His will is not free, and yet he is a powerful man, having shaped the lives of those around him in profound and lasting ways. His children, his students, his wife, his brothers and sisters, his colleagues on the road--all will tell you how dramatic George's influence has been. George may be nonautonomous, but he nevertheless exercises tremendous power over others, and he does it for their good.

I want to make it clear that when I deny that women like Felicité and men like George have autonomy I am not asserting that they are inferior. To lack autonomy constitutes no reason to downgrade a person's value. The problem here is that we are trained to interpret "nonautonomy" as a negative judgment about someone's character when autonomy, in this context, should be a descriptive rather than a normative term. No one would accept a definition of autonomy according to which they did not qualify as autonomous. But if in principle no one can fall outside our definition, then the concept is useless. In order to presume that most adults are autonomous, we must be at least willing to grant that some are not autonomous.

So, George is weakly nonautonomous. But if Frey is right that the way is open to killing and eating nonautonomous beings, then we would be justified in killing and eating George. That seems wildly counterintuitive.

Frey might try to save his thesis by denying one of two things about George. He might try to deny that he is a normal adult human. By putting him in the class of severely brain damaged infants--and cows--he could simply assert that I have not chosen a typical human being as my example. This response is very weak. If George is not a normal adult human then I do not know one. We may safely assume that Frey will not try this route of escape.

More plausibly, Frey could try to deny that George lives a life of high moral quality. Such an argument might go as follows. While George has many wonderful qualities and is certainly a normal human, his life would nonetheless be better, morally speaking, if he were to exercise more of his free will. By leaving his job and becoming a golfer he could continue to exercise his fathering and nurturing skills but in an environment he had chosen for himself. On this interpretation, George would not qualify as a counterexample to Frey's view at all. Instead, he would serve to reinforce the importance of autonomy as a measure of morality, being one more example of the truth of the claim that a life with less autonomy is of lower quality than a life with more of it.

But this response begs the question. We could only determine that George's life was inferior *because* nonautonomous if we already knew that a nonautonomous life was *by definition* inferior. Whether one can have moral standing and be nonautonomous is precisely the question we have set out to answer. We cannot justifiably answer it by reformulating it as an assertion.

Both of these descriptions are true:

(1) George has moral standing.

and

(2) George is nonautonomous.

Because (2) is true, George falls below the line Frey has drawn. That shows the irrelevance of autonomy as the line for deciding whether or not the way is open to killing and eating beings who do not measure up to it.

Let us now apply this view of animals to the case of transgenics.

3. Should we genetically engineer hogs?

Transgenic animals are animals into whose DNA humans have inserted a foreign gene, a gene from a source other than the animal's natural parents. The first transgenic mammal was produced by Palmiter in 1982, who injected a growth hormone gene from a rat into the chromosome of a mouse.[38] The resultant animal expressed the rat gene and quickly grew to twice its parents' size.[39] Mice have served as the transgenic mammalian species of choice. By introducing an activated oncogene sequence taken from humans, for example, scientists have produced transgenic mice with an increased propensity for developing neoplasms. The resultant mouse is a scientific model of human disease, fit for experimental inquiry. Mice are chosen because they are extensively studied warm blooded mammals with extensive physiological and genetic similarities to humans; because they reproduce quickly; because they are relatively inexpensive and easier to handle than larger animals; and because there is little public resistance to using them in research.

Scientists are becoming adept at manipulating mice molecular structures. The oncomouse, genetically modified so as to develop malignant tumors, is not atypical. Labs around the world possess a variety of mice that possess susceptibility to diseases of scientific interest. The mice have been produced either through natural or chemically induced mutation, or through genetic engineering. *Life* magazine presented photographs of some of them in a 1995 issue.[40]

> C3HeB/FeJ, "known as the shiverer, has a condition similar to MS [multiple sclerosis]. A genetically caused deficiency in the myelin protein that sheathes nerve cells makes the mouse tremble whenever it tries to move. "
>
> NOR2/LtDn is "blind from a defect in its optic nerve" and "is used to hunt for the genes causing cataracts, glaucoma and retinitis pigmentosa. "
>
> c57BL/6J, "called a tubby . . . has an abnormal fat-triggering gene. Recently, scientists ... found that the gene makes a hormone called leptin, which may cause excessive weight gain, America's most common disease. "

> WLHR/Le "begins to lose its coat 10 days after birth. Scientists hope studying [the mouse] may yield clues to rare forms of hair loss in humans. "

Mice and rats are the preferred species for genetic research and testing because they are small, easy to handle, breed, and house. Mice grow and mature rapidly, and a female mouse will produce a dozen baby mice every three months. The natural life-span of a mouse is only three years, making it relatively easy to study the course of a disease from start to finish. And the physiological system of the mouse is massively similar to that of the human.[41]

In addition to using rodents, researchers are exploring the possibilities of using larger mammals to produce food and pharmaceuticals, and, in a procedure known as xenotransplantation, spare organs for humans. Transgenic farm animals (TFAs) are animals used for food, fiber, pharmaceuticals or organs into whose DNA humans have inserted a foreign gene. Scientists have produced at least one transgenic animal in each of the following species: cattle, sheep, chickens, rabbits, fish, and goats.

The most famous TFAs are doubtless the Beltsville hogs, produced by Dr. Vernon Pursel and colleagues at the U. S. Department of Agriculture Research Station at Beltsville, Maryland, in 1985.[42] Nineteen transgenic swine with human growth hormones lived through birth and into maturity. Experimenters successfully microinjected the piece of DNA encoding the production of human somatotropin into the nucleus of a fertilized pig egg. The extracted embryo was reimplanted into a sow's uterus, the pregnant animal came to term, and the first piglet in history with a human gene was born.

The Beltsville research program was not aimed at producing hogs twice the size of their parents but at producing more cost effective swine, pigs that would convert grain into lean meat faster than their parents while eating proportionately less grain. Such animals would be a boon to certain sectors of the agricultural economy, including most of the pork industry, some hog farmers, and many meat consumers. The industry might cut costs by slaughtering fewer animals per pound of meat; farmers might reduce expenditures on feedgrains while continuing to sell the same amount of pork; and consumers might benefit from industry and farm savings passed on to them at the meat counter.

Nineteen transgenic swine lived through birth and into maturity. Several expressed elevated levels of the growth gene, but none grew more quickly or to greater size than their counterparts in the control group.[43]

However, many suffered from "deleterious pleiotropic effects," medical problems not afflicting the controls.[44] Those animals developed abnormally and exhibited deformed bodies and skulls. Some had swollen legs; others had ulcers, crossed eyes, renal disease, or arthritis.[45] Of 29 founder pigs, 19 expressed either human growth hormone or bovine growth hormone. Among those exhibiting long-term elevated levels of bGH, health was generally poor. Many seemed to suffer from decreased immune function and were susceptible to pneumonia. All were sterile. Later, Pursel would write that "the pigs had a high incidence of gastric ulcers, arthritis, cardiomegaly, dermatitis, and renal disease," concluding that if transgenic swine were to be produced as successfully as transgenic mice, "better control of transgene expression, a different genetic background, or a modified husbandry regimen" would be required.[46]

These TAs and TFAs have obviously been caused to suffer, and those who believe in animal rights may feel a special sense of outrage at the experiments. When showing slides of these mice and hogs to audiences, I have discovered that even those who do not believe in animals rights are unsettled by the lengths to which we have now gone in treating animals like computer desktops, molded to suit our interests.

Are we justified in producing transgenic animals? Our answer will probably turn on our answer to three related questions. Do individual adult nonhuman mammals have interests, in a morally relevant sense? If so, is it *prima facie* wrong, in the same way if not to the same degree, to deprive an animal of living conditions in which its basic biological needs can be met, just as it is *prima facie* wrong to deprive humans of living conditions in which their basic biological needs can be met? And, how important to society is efficient production of livestock?

Take the last question first. New and more efficient techniques for the production of market hogs could have substantial economic benefits, including, a national hog population bred to convert feed into meat with great efficiency; hog breeders, farrowers, finishers, and consumers reaping financial benefits from the animals' efficient digestive tracts; and, comparative economic advantages for American farmers facing competition from other countries. Notice that the gain in social utility here is not simply a gain in productivity, but a gain in the efficiency of the use of resources, including human and plant resources as well as animal resources. Depending upon how heavily we weigh such gains, we might believe the gains will outweigh the costs associated with the suffering of the nineteen transgenic swine.

But will they? How do we decide how much weight to assign to the animals' pain?

Start with an easier case. Suppose the experimental animals in question were human beings. Imagine that the only way to achieve the financial gains was to transfer swine growth hormone genes into fertilized human embryos, implant the embryos in women, bring the embryos to term, raise the resultant nineteen children to maturity, and then transfer the children's somatotropin genes back into the swine. Suppose further that the children in question had sickly malformed bodies analogous to the bodies of the Beltsville hogs. Clearly, the social benefits in this case, even if they were dramatic and sustained, could not be permitted to outweigh the costs. Any who would entertain the possibility that the pain and suffering of children may be justified by gains in economic efficiency of pork production is morally callous, or worse. We should not bring children into the world to use as means to economic ends, so experimenting on human embryos, without knowing what effects the procedures will have on the children the embryos will become, is at least irresponsible.

We should not approve a line of reasoning that would justify the production of Beltsville *humans* because of economic gains in agricultural production efficiency. Of course, you may object, the experimental animals are hogs, not humans, and it is not apparent that we owe hogs what we owe humans, namely, the duty not to be treated as economic pawns. By way of response to this objection, I start with the obvious fact that all living things have basic biological needs (BBNs). BBNs vary by species and, perhaps, even by individual. But in all cases, BBNs are needs that must be met if an individual's welfare is not to be thwarted.

What are some typical human BBNs? To be able to ingest sufficient amounts of uncontaminated protein and water without undue pain; to be able to eliminate bodily wastes without wasting half the day doing it; to be able to maintain sufficient psychological equilibrium that we are able to fall asleep at night; to have access to sufficient open space that we can accelerate our heart rates to one hundred odd beats per minute for half an hour three times a week; to possess a backbone and neck muscles strong enough that our heads do not need external support; to have an immune system not vulnerable to common air borne viruses.

If we are born with a medical condition that deprives us of the ability to have one of our basic needs met, we are the worse-off for it, but we cannot necessarily say someone has harmed us. If, on the other hand, our unfortunate condition is the result of someone's having injured or deprived us, or having injured the fertilized egg we once were, then the offending person has harmed us, done us a moral wrong.

Call this principle (1):

(1) It is (morally) wrong to deprive an individual, *S*, of something they must have if their BBNs are to be met.

Notice that it is *S* who is wronged by the offending action, and not someone or something else. This means that the principle can only apply to beings with a welfare that may be promoted or harmed. There are, of course, many things in the world without welfares, and such things cannot be directly harmed. Examples include natural objects like mountains and piles of sand, and human artifacts like bridges and computer printers. You might harm the owner or user of these things by mishandling the object, but you cannot harm the *object,* because natural objects and artifacts do not have a good of *their* own. So (1) does not apply to things, because things are not individuals, do not have biological needs, lack intrinsic value, and have no good or welfare of their own.

Individuals are animated beings, beings that exhibit goal-directed behavior in which the goal or principle of movement is internal to the being. Humans are individuals, but fingers are not; hogs are individuals, but a serum with hog growth hormone in it is not; tomato plants are individuals, but their fruit is not.

Here an obvious problem with (1) surfaces. If the principle were true, we could not justifiably sever the head of a cabbage from its root in order to eat it. So doing would deprive an individual *S,* the cabbage, of something *S* needs in order to have its basic biological needs met.[47] I trust our common intuition here, that killing cabbages in order to feed ourselves is morally permissible and, more generally, that there are many individuals, including all onions and cabbages, toward which we do not possess even a *prima facie* duty not to deprive them of things they need to have their BBNs met.

What distinguishes individuals that may be killed from individuals who may not be killed? Previous arguments in this chapter point toward this answer: *having a future,* meaning the capacity to take an interest in, and to accomplish, things yet undone. I understand "the capacity to take an interest" in the way many others have: *S* has the capacity to take an interest in *X* if and only if *S* has feelings of well-being that may be affected by *X*.[48] Obviously, cabbages are not conscious in this sense, because they lack feelings. Cabbages lack feelings because they lack the hardware necessary to have feelings, namely, a brain, central nervous system and sensory receptors. Lacking feelings, they lack the capacity to take an interest in things in their future, or even to have a future.

Thus, individual *A* in the plant kingdom may justifiably be deprived of something it must have if its BBNs are to be met because, even though that individual has BBNs, it does not have consciousness. Having no future, it cannot be harmed by depriving it of a future.[49]

We must amend (1), therefore, to accommodate the claim that it is not always *prima facie* morally wrong to kill individuals. I offer, then, (2):

> **(2) It is *prima facie* (morally) wrong to deprive a conscious individual with a future of the things it must have if its BBNs are to be met.**

Combining (2) with

> **(3) The Beltsville experiments deprived individual hogs of things they need to have their BBNs met,**

and (2) and (3) with (4):

> **(4) Hogs are conscious individuals with futures,**

we arrive at this conclusion (5):

> **(5) It was *prima facie* (morally) wrong to deprive the Beltsville hogs of the things they needed to have their BBNs met.**

Some will want to contest (4), and argue that pigs either are not conscious, or do not have a future, or both. If pigs are more like computer desktops or bridges than they are like children, then we can no more harm a pig by unintentionally breeding it to have a bad body than we can harm a bridge by unintentionally designing it to lack earthquake sustaining power.

But pigs are clearly different from bridges, not only because they can move themselves around, experience pain and pleasure, and lead social lives, but because there are things pigs must have in order to have their BBNs met. If a pig's bone structure is unable to bear its weight; if its sensory systems are unable to give it reliable information; if its immune functions fail to protect it from common diseases; then the pig will lead a deprived life, unable to engage in the goal directed behaviors characteristic of its species. It will, variously, not be able to eat or mate or root or play with or care for its young or establish a social order or investigate its environment. The pig itself will fare poorly if it does not have the things it needs to have its BBNs met. If scientists engineer pig embryos that develop into individuals with deformed bodies or poorly developed brains, they have harmed the pig. Whether they do this intentionally or unintentionally should bear on how much culpability we assign to the scientists, but it should not affect the question of whether the pigs themselves have been harmed. So this objection to (4) fails.

Again, one might grant that pigs are individuals and have BBNs, but insist that pigs are unable to take an interest in anything that may affect their future well-being. If so, then they, like cabbages, are not *conscious* individuals, and so cannot have a right not to be deprived of things they need to have their BBNs met. I believe this criticism is wrong, and I will argue that pigs are able to take an interest in some things. But I want to avoid the language of animal rights, because the tradition of rights talk is inimical to the sort of moral attitudes I wish to encourage. Rights talk encourages us to think of the moral sphere as an arena of atomistic units warring with each other to defend turf against invaders. I want to encourage views of the moral sphere in which individuals are construed more interdependently, engaged in projects that are more cooperative than competitive.

I have a difficult problem in avoiding the individualism of rights language without tearing down the legal and philosophical fence around individuals which rights language has so admirably erected. I think the way to do this is to try to show that the notion that "it is wrong to deprive a being that can take an interest in having its BBNs met of the things it needs to have its BBNs met" is a *primitive notion.* That is, while you can give examples of the wrong that obtains when an individual is deprived, you can do nothing more by the way of giving reasons that it is wrong than telling stories. There is no further justification that can be offered for why it is wrong to deprive an individual but, fortunately, there is no further justification needed.

Ethical reflection means giving reasons for our judgments. When we say some action is wrong, others are justified in asking us why we think that. When we give a reason, that reason may be formulated as a general moral principle. But our partners may want to know why that principle is true, and may justifiably ask us to ground our reasons for our decisions in some more basic, ultimately vindicating, reason. The work of ethics proceeds this way, with claims being grounded in reasons, and reasons in principles, and principles in theories.

But the dialectic of ethics does not go on forever; at some point we reach the ultimately vindicating ground of our reasoning. When we reach this ground, others will ask us why we rest on that ground, and we may be tempted to try to provide a reason. We should resist this temptation, because, if we have truly reached bedrock, there is nothing further for us to say. Ludwig Wittgenstein once remarked that the most difficult part of justification in philosophy is to recognize a justification *as* a justification, and to stop.

Actions which are properly categorized as wanton destruction of innocent humans for trivial reasons is such a stopping place. Assuming that we have a true case of such heinousness, we need not think we must offer further justification for *why* we believe a heinous action to be a stopping place. Here is a true case: riddling my nephew with bullets merely to try out the recalibrated sight on my automatic weapon. Such actions are always and irredeemably evil, and we need not consider the arguments of someone who wants to try to argue that this act is morally justifiable.

We must, of course, listen to arguments that try to justify the killing of humans for non-trivial reasons, such as in cases of self-defense, war, and the punishment of criminals. If I am shooting my nephew out of self-defense, that is another case than the one described above and open, perhaps, for potential justification. But we need not try to reason with the person who, simply for the sake of argument, wants in bad faith to defend truly wanton cruelty.

The question before us is whether the killing of animals for food qualifies as a stopping place. Clearly not. We must take seriously the arguments of those who think we are justified in eating animals. My point is only this. As we reason about vegetarianism, we need not reason about the justification of trivial killing of humans.

My claim is not that the obligation not to kill animals for food is unarguable and on all fours with the obligation not to kill innocent humans for trivial reasons. My claim is that this obligation may be overridden by other obligations, but it may not be *lightly* overridden. A preference-interest for the taste of meat when other sources of nutrition are easily and cheaply available, is not in my judgment a weighty preference-interest.

Some hold that killing for trivial reasons is not wrong if it involves killing a being that has only brief, short-term, desires. Ruth Cigman, for example, holds that killing is wrong only insofar as the victim is capable of having what she calls "categorical desires," desires in which the victim is not "blindly clinging on to life," but in which it also "possesses the related concepts of long-term future possibilities, of life itself as an object of value, of consciousness, agency and their annihilation, and of tragedy and similar misfortunes. "[50] Humans are able to have these sophisticated concepts and desires, and death harms them by depriving them of their categorical desires.

I have argued that the mere having of desires is sufficient to establish a moral right not to be blocked for trivial reasons from pursuing those desires. It does not matter whether the desires in question are long-term, categorical, desires, such as wanting to see one's daughter graduate, or short-term, humble desires, such as wanting to continue stroking a baby's hair as she falls asleep. We have a basic moral right not to have others interfere with our preference-interests, basic or trivial, so long as their satisfaction does not conflict with the welfare of another desiring creature.

We can now articulate more carefully the wrongness in killing a cow to eat it when our basic interests do not depend on it. In killing a cow, we deprive it of the ability to pursue whatever is its current preference-interest. We deprive it of the ability to do what it wants to do in the future, say, to finish chewing its cud or to cross to the other side of the pasture to drink. In killing mammals, we deprive them of their future, of their ability to finish doing whatever they now want to do, say, stroking their offspring's hair as she falls asleep.

The reason the Beltsville hogs were tampered with at the embryonic stage was to produce brave new pigs that would grow more quickly to slaughter weight, and the purpose of much TFA research is to produce animals to be killed for meat. What is wrong with this research is not that it involves gene splicing but, rather, that it is aimed at morally objectionable goals.

I can tell you why I think hogs are individuals, and I can tell you why I think harming innocent individual humans by depriving them of the basic things they need is morally wrong, and I can give you examples of cases I think involve wanton harm to humans and I can tell you why I think hogs are innocent in ways analogous to humans. But I cannot tell you why I think harming innocent humans for trivial reasons is morally wrong. If you tell me that you see no moral wrong in wantonly harming humans, I have three responses open to me. I can first ask you if you are serious, and try to decide whether I think you are speaking in good faith. If I decide you are serious, I can, second, tell you stories of deprivation. If after several rounds of stories, told in increasingly graphic detail, I decide that you are still serious, then what option do I have other than to worry that you may not have been brought up in the right way, and that you may be dangerous?

My claim is that the following idea is a primitive notion, which reasonable persons, once they understand it, must accept it: "To harm an individual human by depriving it of things it needs to have its BBNs met is *prima facie* morally wrong." If I am correct, then we can use the idea of primitive notions in place of rights talk. That is, wherever philosophers have justifiably ascribed a basic "right" to *r* of some individual, I want to say that reasonable persons brought up correctly, who understand what it would mean to deprive the individual of *r,* also understand and accept the primitive notion that "depriving the individual of *r* is *prima facie* morally wrong." And that is all we can say, or need to say, about the matter.

If the argument about primitive notions works, we can offer strong protections for individuals without recourse to rights talk. And that would allow us to show why we have strong duties not to deprive conscious beings of things they need to have their BBNs met. As Tooley has shown, you cannot have a right to something unless you are capable of taking an interest in it.[51] You cannot take an interest in something unless you are conscious. Thus, only conscious beings, a subset of the class of all individuals, are candidates for the strong protections traditionally formulated in rights language. So it remains to say why we should extend this primitive moral notion to other vertebrates.

My reason for thinking hogs can take an interest in something is the same as my reason for thinking my four-month old daughter can take an interest in something. In my daughter's case, my belief is based on inferences drawn from observations of her behavior. I remember watching Krista's eyes follow a mobile slowly turning over her crib. Her lids would open slowly after her nap, wander around the room, and then fix on the motions above her. I surmised she was "taking an interest" in the mobile because her eyes would sometimes stray toward me, but she would shut out the distraction, even as I strained to get her attention, focusing once again on the revolving colors.

In the hog's case, my belief is based on similar inferences. I say to my uncle, an Iowa farmer, "That old sow really took an interest in the tire we threw in there." When I say "took an interest," I mean it in exactly the same sense as when I apply it to my daughter. Consider the behavioral signals each gives; the level of visual and mental focusing going on; and the kinds and grounds of inferences I make on the basis of those signals. All of these things are identical in the two cases. I see the hog's eyes open as the tire sails in; I see the animal slowly rise to face the foreign object; I watch as he cautiously approaches it, snorting and backing at irregular intervals. I surmise, as I watch him spend the rest of the morning intently nosing the tire treads, oblivious to me and to his pen mates, that his attention has been captured by the tire. What more simple or elegant or efficient explanation is there than to say, "the pig has taken an interest in" the object?

Being the crux of the argument, this point bears underscoring with other examples. Boars can take a monogamous interest in a single gilt coming into heat; in a knot hole knocked through a pine board; in the bristles of another pig's back. They can take an interest in people, and they can, if the Hollywood animal trainer Frank Inn is to be believed, take an interest in ignoring people. Inn reportedly said:

> You can force a dog, a chimp or a horse to do something, but a pig, no. Pigs won't take punishment. Reprimanding will work

> with a dog, but with a pig, never. If you reprimand a pig he won't like you, won't respond to you and won't even take food from you. You can see temper in pigs. If I scold them, they scold right back.[52]

If an individual has the capacity to take an interest in something, it must be capable of losing interest in something, too. We usually lose interest when we become bored, when the thing occupying our attention no longer intrigues us. That happens when something that once intrigued us no longer presents new opportunities or facets to our imagination. Our imaginations, of course, are not infinitely plastic, and the things that will continue to engage our fancy over a long period of time fall within limits drawn by our genetic background, social upbringing, and professional training.

Pigs can lose interest in things. In one experiment, hogs were trained to carry coins from one end of their pen to the other and to deposit them in a bank. Researchers found that the animals quickly progressed to a stage where the animals would carry four or five coins before needing reinforcement. As they put it, "pigs condition very rapidly" or, as we might more accurately put it, pigs have a tremendous capacity for becoming interested in things.

Being intelligent, pigs also have a high threshold of boredom. Unless a new object or behaviorial stimulus has some relationship to the basic wants and drives of the animal, we might predict that the pig's interest will wane. In a development that could only have surprised scientists committed to a behaviorist paradigm, that is what happened in this case. After a period of several weeks, the experimental animals stopped performing the chore they had been "conditioned" to do.

> This particular problem behavior developed in pig after pig, usually after a period of weeks or months, getting worse every day. At first the pig would eagerly pick up one dollar, carry it to the bank, run back, get another, carry it rapidly and neatly, and so on, until the ratio was complete. Thereafter, over a period of weeks the behavior would become slower and slower. He might run over eagerly for each dollar, but on the way back, instead of carrying the dollar and depositing it simply and cleanly, he would repeatedly drop it, root it, drop it again, root it along the way, pick it up, toss it up in the air, drop it, root it some more, and so on.[53]

The researchers described the pig's actions as "problem behavior" resulting from a breakdown in "conditioning." We might describe it more accurately by calling it a natural loss of interest in objects and stimuli not consistent with the pig's basic biological needs and wants. The pigs, being pigs, were more interested in rooting the coins than in putting them in the bank.

The behaviorists' conclusion bears citing:

> We thought this behavior might simply be the dilly-dallying of an animal on a low drive. However, the behavior persisted and gained in strength in spite of a severely increased drive--he finally went through the ratios so slowly that he did not get enough to eat in the course of a day. Finally it would take the pig about 10 minutes to transport four coins a distance of about 6 feet. This problem behavior developed repeatedly in successive pigs . . . (We concluded] that these particular behaviors to which the animals drift are clear-cut examples of instinctive behaviors having to do with the natural food getting behaviors of the particular species.[54]

To call the pigs' behavior "instinctive" begs the question whether the animals are beings with mental powers comparable to those of, say, a human two-year old. An alternative explanation is that the hogs' behaviors were clear-cut examples of this species' ability to take an interest in, and then to lose interest in, novel environmental conditions. In saying this, we can use the phrase, "take and lose interest in," in exactly the same sense as when we apply it to very young children.

My two-year old "has a future" because she takes an interest in things and can interact with her world so as to shape it to her desires. She can learn. She can figure out how to get things she wants. Pigs can similarly take an interest in things, interact with their world so as to shape it to their desires, learn, and figure out how to get things they want. A pig, in sum, has a future. They not only have BBNs; they also have welfare, goods of their own that may be promoted or thwarted. It is a primitive notion that it would be morally wrong to harm a human infant by depriving it of the things it needs to have its BBNs met. Is it not also a primitive notion that it is morally wrong to harm a pig by depriving it of things it needs to have its BBNs met?

There is a difference between killing an animal for a trivial reason and killing it for a good reason. Suppose that a scientist wanted to do research that would produce transgenic animals in order to save human lives. Should we allow such an experiment?

Let us once again begin with an easier case. Suppose a scientist wanted to produce transgenic humans. How would we respond?

Imagine that a woman's genetic heritage makes her a high risk to develop cancer at an early age. Her husband is also a high risk. Knowing full well that any children they bring into the world will almost certainly be saddled with a genetic predisposition to develop malignant tumors early in life, the couple still cannot overcome their desire to have a child of their own.

Now suppose that science has progressed to the point that medical researchers feel confident that they can insert a gene into the woman's ova that will dramatically reduce the risks of cancer for the child. Suppose further that, due to a combination of regulatory hurdles and technological shortcomings, the researchers can only access the gene from another species, say, the ape. The baby we are now envisioning is the first transgenic human. What responsibilities would scientists have to her? What moral rules ought to guide us as we take the first tentative steps down the path of human germ cell therapy?

I will not try to develop a complete list of rules and regulations about this complex subject here. The medical community is now beginning to think about the more fundamental question, whether to allow the insertion of foreign genes into human sex cells at all and, as Paul Thompson points out, there presently seems to be "a widely shared conviction that human eugenics is morally wrong."[55] But if the consensus on that issue turns out the way it has with regard to the insertion of foreign genes into animal sex cells, then we shall soon have to begin devising such guidelines, because the option of foregoing all germ cell therapy will not be a live one. Presuming that, one day, we will have transgenic human production, what basic rules ought to bind us?

(a) No harvest THs.

Harvest animals are animals intentionally bred and raised for the purpose of being killed at a young age. In our culture, harvest animals fall largely into one of two groups. First, there are experimental animals, primarily mice and rats, which are killed so that researchers may do autopsies and learn scientific information. Second, there are farm animals, primarily chickens, cows, and hogs, which are slaughtered for their meat. Harvest transgenic *humans* would be transgenic humans intentionally bred and raised for the purpose of being harvested at a young age. I cannot imagine anyone proposing to raise humans for meat, but it is not implausible to imagine someone in the future proposing to bring a handful of injected human ova to term in order to discover whether the injected genotypic change will be expressed phenotypically. The argument, of course, would be that hundreds of thousands of humans would eventually benefit from the harvest THs. But I have great trust in our intuition here, that we should not allow the production of experimental humans-for-slaughter, no matter how many other humans might be saved.

Doctors and scientists should protect the basic interests of all human subjects used experimentally, but a special obligation exists to protect innocents. Not all writers are as uncompromising on this point as Hans Jonas, but the vast majority would agree with the spirit of his remark on the morality of using an unconscious or subconscious patient in research:

> Drafting him for non-therapeutic experiments is simply and unqualifiedly impermissible; progress or not, he must never be used, on the inflexible principle that utter helplessness demands utter protection.[56]

Suppose that the happy parents of the low cancer risk TH infant agree to let their doctors conduct a certain number of nontherapeutic tests on their child. They understand that the baby will not be harmed by these tests and, indeed, the youngster grows up to be healthy and content. After fifteen years, however, the adolescent decides that enough is enough, and makes her wish known that the tests end. Her refusal to grant consent should be treated the same as anyone's refusal to grant consent, just as any informed choice of a TH should be treated in the same way that we would treat the informed choice of a non-TH. The classic legal principle of informed consent was stated by Chief Judge Cardozo:

> Every human being of adult years and sound mind has a right to determine what shall be done with his own body; and a surgeon who performs an operation without his patient's consent, commits an assault, for which he is liable in damages.[57]

The TH I have been imagining is one that is well positioned to give consent. But if we want to protect her, how much more we should want to protect a TH who turns out not to be so well positioned. Suppose the experiment, tragically, went awry, and the resultant child never developed the mental capacities required to give informed consent. I believe we should not run any nontherapeutic tests on such a misfortunate, simply because people who are least prepared to give informed consent, or who are utterly unable to give informed consent, should be most protected against experiments and tests that are not undertaken for *their* well being.

Which THs would we ideally use as experimental subjects? Those best able to understand and bear the risks to which they would be submitting themselves, and who would be most disposed and prepared to care for their TH offspring in the event that something went wrong. Jonas' way of talking about informed consent is apt. Samuel Gorovitz summarizes it as follows:

> Morally permissible use of human beings in medical experimentation requires that they be those persons with a maximum of identification, understanding, and spontaneity--the most highly motivated, the most highly educated, and the least 'captive' members of the community.[58]

Notice that nothing I have said prohibits the production of transgenic humans. If a transgenic procedure would make a future human being better off (by, say, removing a gene for cystic fibrosis), and if science could benefit from studying the future individual, I see no obvious reason why that person might not also be the subject of future testing, providing that certain conditions were met. One condition would be that the testing itself would not harm the person. Another would be that the person's informed consent would be required. If, for example, scientists simply wanted to observe the TH to find out if the targeted gene had actually been deleted, and if they could make their observations without harming the subject or infringing on her informed consent, then doing so would not be impermissible according to ***(a)***.

(b) No worse-off THs.

It would also be objectionable to experiment on THs, even with the informed consent of the TH, if the experiment would seriously undermine the well being of the TH. Claude Bernard, a leading nineteenth century physician, wrote that the very foundation of medical morality is "never to carry out on a human being an experiment that cannot but be injurious to him to some degree, even if the outcome could be of great interest to science, that is to say, the health of other human beings."[59] Following Bernard's principle, we should not inject foreign genes into human ova if we have good reasons to suspect that the life of the prospective TH will be worse-off than it would have been had it not been tampered with at the embryonic stage.

There are many things you can do to me without making me worse-off, because my well being is not measured by a single criterion, such as the absence of physical disease. As welfare is a composite measure of many different variables including one's own feelings, a slight setback in one area may sometimes be overcome by gains in another. For example, a patient dying of lung cancer might feel better off than an overworked single mother, depending upon how each person feels about her situation. If the single mother is under financial and emotional stress and constantly battles depression, she may have lower feelings of well being than the elderly woman who has spiritually and enthusiastically embraced her fate. Assessing welfare is a difficult chore.

But not an impossible one. There are many things I can do to you that will clearly make you worse-off, and there are many things you can do to me that will clearly may sometimes make me better off. To distinguish clear harms and benefits from the vast grey areas that lie in-between them, it is important to draw attention to our fundamental interests, to things we must have.

Some activities are pleasant, but we are not entitled to them. I would be worse-off without income sufficient to pay for violin lessons; without leisure time to spend with my brother-in-laws at the movies; without an indoor basketball court in which to practice my fifteen-footer. I take an interest in these activities, and they are good for me, but if my violin money, movie time, and gym privileges were taken away, I could flourish nonetheless by substituting different interests. Things in which we take an interest but to which we have no moral right, are nonbasic interests (NBIs).

A public policy that deprives me of one of my NBIs, say, the ability to play the violin, will not necessarily make me worse-off. What will necessarily make me worse-off is a transgenic procedure that deprives me of one of my basic interests (BIs), such as, my ability to make or hear sounds.

I have argued previously for two rules regarding the production of transgenic humans. Given the strength of the animal rights theory, the applicability of these two rules to the production of transgenic animals is straightforward.

The rules governing the production of transgenic humans should govern the production of transgenic animals.

No worse-off transgenic animals.

No harvest transgenic animals.

According to these rules, we have already gone too far in making transgenic animals, and we should not allow further experiments such as those that produced the Beltsville hogs and hairless mice.

Notes

1. William Hedgepeth, *The Hog Book* (New York: Doubleday, 1978), pp. 54-55.
2 Tom Regan, *The Case for Animal Rights* (Berkeley: University of California Press, 1983).
3. Cf. Kathy and Bob Kellogg, *Raising Pigs Successfully* (Charlotte, VT: Williamson, 1985), p. 23.
4. Michael Tooley, discussing the alleged right to life of fetuses, gives what he calls the "particular-interests principle," which draws on Joel Feinberg's general "interest-principle." Tooley's principle states "It is a conceptual truth that an entity cannot have a particular right, R, unless it is at least capable of having some interest, I, which is furthered by its having right R." Michael Tooley, "In Defense of Abortion and Infanticide," in Joel Feinberg, ed., *The Problem of Abortion*, 2nd ed. (Belmont: Wadsworth, 1984), p. 125. Tooley's example of the principle is that "an entity cannot have a right to life unless it is capable of having an interest in its own continued existence" (p. 132).
5 Tom Regan, "The Case for Animal Rights," in Peter Singer, ed., *In Defense of Animals* (NY: Basil Blackwell, 1985), p. 13.
6. Cf. Hedgepeth's informal survey of the meanings of hog sounds, like *groonk*, *rah*, *wheenk*, and *Wheeeeeeeiiiiiii* (p. 137).
7. Paul Taylor, *Respect for Nature*: *A Theory of Environmental Ethics* (Princeton: Princeton University Press, 1986), p. 81.
8. Taylor, p. 83.
9. Taylor, p. 278.
10. Taylor, p. 283.
11. Taylor, p. 55.
12. See, for example, Callicott, *In Defense of the Land Ethic: Essays in Environmental Philosophy* (Albany: S.U.N.Y. Press, 1989), esp. pp. 33-36; and Midgley, *Animals and Why They Matter* (Athens: U. of Georgia Press, 1983), esp. pp. 112-24.
13. Much of the following information is taken directly from a study by E. S. E. Hafez and J. P. Signoret, "The Behaviour of Swine," in Hafez, ed. *The Behaviour of Domestic Animals* (Baltimore: Williams and Wilkins, 1969), pp. 349-390.
14. On this point, Klaus Immelman argues that it is very difficult to say that a specific behavioral trait has been caused by the process of domestication. Immelman, *Introduction to Ethology*, tr. Erich Klinghammer (New York: Plenum, 1980), p. 197. The reason is that changes in behavior can result from the animals' responses to environmental conditions. Such changes would have to be considered modifications made by the animal itself, not by human selection of specific genetic arrangements.
15. Taylor refrains from saying food animals have a right to life. At one point he seems to imply the history of domestication has reduced food animals from the status of beings deserving respect to the status of "machines, buildings, tools, and other human artifacts" (p. 56). But he immediately adds we must treat them in ways that are good *for them*, and "this is a matter that is quite independent of whatever usefulness [they] might have to humans" (p. 57). He writes, further, that "the question of how [domestic animals] ought to be treated could not be decided simply by seeing what sort of treatment of them most effectively brings about the

human benefit for which they are being used" (p. 57). Taylor's view is not unambiguous, but this is because his primary concern is with our duties toward wild living things and not with our duties toward domesticated animals.

16 J. Baird Callicott, "Animal Liberation and Environmental Ethics," in *In Defense of the Land Ethic*, pp. 55-56.

17 Callicott, Land Ethic, p. 56.

18. J. Baird Callicott, "Animal Liberation and Environmental Ethics: Back Together Again," *Between the Species* 4 (Summer 1988): 167.

19. Taylor, p. 295.

20. Taylor, p. 296. The reference for Taylor's figures is Francis Moore Lappe, *Diet for a Small Planet* (New York: Ballantine, 1971).

21. R. G. Frey, *Interests and Rights: The Case Against Animals* (Oxford: Clarendon Press, 1980), especially chapters 2 and 3. Frey takes the phrase "marginal cases" from Jan Narveson, "Animal Rights," *Canadian Journal of Philosophy* 7 (1977): 167.

22. For his arguments against the acceptability of the notion of moral rights, see *Interests*, pp. 7-17, and Part III (pp. 43-98) of *Rights, Killing and Suffering: Moral Vegetarianism and Applied Ethics* (Oxford: Basil Blackwell, 1983). For his concessions about the usefulness of the notion, see ch. 9 of *Rights*, "Rights, Their Nature, and the Problem of Strength," pp. 67-82.

23. *The Monist* 70 (January 1987): 49-63.

24. Frey uses the phrase "normal adult humans" in his first sentence. "Autonomy," p. 50.

25. "Autonomy," p. 50. It is worth noticing that the concept of moral rights is also central to the theories he identifies.

26. Regan, *The Case for Animal Rights* (Berkeley: University of California Press, 1983), pp. 84-85. Cited in Frey, p. 60.

27. Frey, p. 51.

28. "Autonomy," p. 53.

29. "Autonomy," p. 53.

30. "Autonomy," p. 54.

31. Frey, Autonomy, 58. Cf. Paul Taylor, *Respect for Nature: A Theory of Environmental Ethics* (Princeton: Princeton University Press, 1986).

32. How many people in the world are like George? I am currently reading two realistic novels, Louise Erdrich's *The Beet Queen* (New York: Henry Holt, 1986) and Iris Murdoch's *The Philosopher's Pupil* (New York: Viking, 1983). Out of a total of roughly twenty major characters in these two contemporary works, I would call all twenty "normal adult humans," roughly fifteen of them morally admirable people, but only five of them autonomous in Frey's sense.

33. In "The Value of Autonomy," *The Philosophical Quarterly* 32 (January 1982): 35-44, Robert Young raises the question of whether autonomy is always a virtue.

34. "Autonomy," p. 57.

35. The quotation is from Regan's paper, "The Case for Animal Rights," in Peter Singer, ed., *In Defence of Animals* (Oxford: Blackwell, 1985), p. 22. Quoted in Frey, "Autonomy," p. 59.

36. In what follows, I am drawing on some ideas of Harry Frankfurt's, and a suggestion of Phil Quinn's.

37. Harry Frankfurt, *The Importance of What We Care About* (Cambridge: Cambridge University Press, 1988), p. 21.

38. R. D. Palmiter, R. L. Brinster, R. E. Hammer, M. E. Trumbauer, M. G. Rosenfeld, N. C. Brinberg, and R. M. Evans, "Dramatic Growth of Mice that Develop From Eggs Microinjected with Metallothionein-Growth Hormone Fusion Genes," *Nature* 300 (1982): 611-615; R. D.

Palmiter and R. L. Brinster, "Transgenic Mice," *Cell* 41 (1985): 343-345; and R. D. Palmiter, G. Norstedt, G. E. Gelinas, R. E. Hammer, and R. L. Brinster, *Science* 222 (1983): 809.

39 For transgenic mice, see R. D. Palmiter, R. L. Brinster, R. E. Hammer, M. E. Trumbauer, M. G. Rosenfeld, N. C. Brinberg, and R. M. Evans, "Dramatic Growth of Mice that Develop From Eggs Microinjected with Metal loth ione i n-Growth Hormone Fusion Genes," *Nature* 300 (1982): 611-615; R. D. Palmiter and R. L. Brinster, "Transgenic Mice," *Cell* 41 (1985): 343-345; and R. D. Palmiter, G. Norstedt, G. E. Gelinas, R. E. Hammer, and R. L. Brinster, *Science* 222 (1983): 809. For transgenic swine, see R. E. Hammer, V. G. Pursel, C. E. Rexroad, Jr., R. J. Wall, D. J. Bolt, K. M. Ebert, R. D. Palmiter, and R. L. Brinster, *Nature* 315 (1985): 680-683; C. A. Pinkert, V. G. Pursel, K. F. Miller, R. D. Palmiter, and R. L. Brinster, *Journal of Animal Science* 65 (Suppl. 1, 1987), Abstract; and C. A. Pinkert, "Gene Transfer and the Production of Transgenic Livestock," *Proceedings of the* U. S. *Animal Health Association* (in press): 122-133.

40. Charles Hirshberg, "Altered States," *Life* (Dec 1995): 80-86. Also, see the arresting photographs that accompany the article, by George Steinmetz.

41 Sandra Blakeslee, "Of Mice and Men," in Pines, Maya, ed. "Blazing a Genetic Trail." Bethesda, MD: Howard Hughes Medical Institute, 1991.
www.gene.com/ae/AB/IE/Of_Mice_and_Men.html

42 Vernon G. Pursel, Carl A. Pinkert, Kurt F. Miller, Douglas J. Bolt, Roger G. Campbell, Richard D. Palmiter, Ralph L. Brinster, Robert E. Hammer, "Genetic Engineering of Livestock," *Science* 244 (16 June 1989): 1281-1288.

43 Transgenic refers to animals with "foreign" genes, genes from species with which they cannot breed naturally. The changes in the germ line are accomplished via recombinant DNA techniques.

44 Michael W. Fox, "Genetic Engineering and Animal Welfare," *Applied Animal Behaviour Science* 22 (1989), p. 107. Pleiotropism is a biological term meaning "multiple." In this context, it refers to the many mechanisms involved in the genes' control of the physical make-up of the animal.

45 Of 29 founder pigs, 19 expressed either human growth hormone or bovine growth hormone. Among those exhibiting long-term elevated levels of bGH, health was generally poor: "the pigs had a high incidence of gastric ulcers, arthritis, cardiomegaly, dermatitis, and renal disease," Pursel (1989): 1281.

46 Pursel (1989): 1281.

47 For an argument that plants are individuals, have needs, and biological interests, and that this qualifies them for direct moral consideration, see Gary E. Varner, "Biological Functions and Biological Interests," *The Southern Journal of Philosophy* 28 (1990): 251-270.

48 Cf. S. F. Sapontzis, *Morals, Reason, and Animals* (Philadelphia: Temple University Press, 1987), p. 74. Sapontzis credits Leonard Nelson for this analysis. See Nelson, *A System of Ethics* (New Haven: Yale University Press, 1956), pp. 136-144.

49 However, I believe that there are other grounds, of an environmental sort, that can provide reasons that it is *prima facie* wrong to kill plants. These environmental grounds show that some significant good is harmed when plants are killed willy-nilly, without good reason. But that good does not include the welfare of the plant itself.

50. Ruth Cigman, "Death, Misfortune and Species Inequality," *Philosophy and Public Affairs* 10 (Winter 1980): 59.

51 Discussing the alleged right to life of fetuses, Tooley defends what he calls the "particular-interests principle," which draws on Joel Feinberg's general "interest-principle." Tooley's principle states "It is a conceptual truth that .an entity cannot have a particular right,

R, unless it is at least capable of having some interest, I, which is furthered by its having right R." Michael Tooley, "In Defense of Abortion and Infanticide," in Joel Feinberg, ed., *The Problem of Abortion,* 2nd. ed. (Belmont, CA: Wadsworth, 1984), p. 125. Tooley's example of the principle is that "an entity cannot have a right to life unless it is capable of having an interest in its own continued existence" (p. 132).

52 Quoted in Hedgepeth (1978), 111.

53 Keller Breland and Marian Breland, "The Misbehavior of Organisms," *American Pscyhologist* 16 (1961): 683, reprinted in Robert W. Hendersen, ed., *Learning in Animals* (Stroudsburg, PA: Hutchinson Ross, 1982), p. 286.

54 Breland and Breland, p. 683.

55 Paul Thompson, "Designing Animals: Ethical Issues for Genetic Engineers," *Journal of Dairy Science*, 75 (1992): 2296.

56. Hans Jonas, "Philosophical Reflections on Experimenting with Human Subjects," *Daedalus* (Spring 1969), *Ethical Aspects of Experimentation with Human Subjects*, and partially reprinted in Samuel Gorovitz, et al., *Moral Problems in Medicine* 2nd ed. (Englewood Cliffs: Prentice-Hall, 1983), p. 114.

57. *Schloendorff* v. *New York Hospital* 105 N.E. 92; N.Y. 1914). Cited in Alan Donagan, "Informed Consent in Therapy and Experimentation," *The Journal of Medicine & Philosophy* 2 (1977): 310-327; partially reprinted in Gorovitz, p. 162.

58. Gorovitz's explanation of Jonas' position appears in a footnote in Gorovitz, p. 114.

59. Quoted by Donagan, in Gorovitz, p. 165. Donagan gives the following references: Quoted in R. A. McCance, "The Practice of Experimental Medicine," *Proceedings of the Royal Society of Medicine* 44 (1950): 189-194, in Irving Ladimer and Roger W. Newman, *Clinical Investigation in Medicine* (Boston: Law-Medicine Research Institute, Boston University, 1963), pp. 48-57.

Chapter 4

Against Ag Biotech (1994)

> *The secret workings of nature do not reveal themselves to one who simply contemplates the natural flow of events. It is when nature is tormented by art, when man interferes with nature, vexes nature, tries to make her do what he wants, not what she wants, that he begins to understand how she works and may hope to learn how to control her. . . . It is my intention to bind, and place at your command, nature . . .*
>
> - Francis Bacon (according to Farrington) [1]

> *Humanity cannot afford to acknowledge all of the blood that it spills and the destruction it inflicts on the world in its effort to perpetuate itself . . . [and to place nature] under our control . . .*
>
> - Jeremy Rifkin [2]

When Francis Bacon declared his intent to torment and interfere with nature, he probably did not envision sickly experimental hogs with human genes. But the Baconian desire to understand nature and place "her" at our command has entrenched itself in our collective psyche, and the bioengineering epoch has enabled us to impose our desires in ways Bacon could not have imagined. In so doing, have we stepped over the bounds of decency?

Many think not. According to traditional morality, animal suffering may be justified if the results are likely to benefit humans. In the case of transgenic animals, knowledge gained from such experiments were instrumental in discovering ways to improve medicine. For example, pharmaceutical proteins, including human factor IX, blood clotting factor, and alpha-1 antitrypsin, are now secreted in the milk of transgenic sheep, producing a purer and cheaper source of these proteins. To bring a higher quality of life for some humans is impossible without the use of animals in scientific research. Andrew Scott praises the level of creative effort involved in such efforts, asserting that gene-splicing has lifted us into the pantheon:

> Mankind is undoubtedly the pinnacle of evolution's achievements so far . . . As molecular biology unlocks the secrets of how life manages to live . . . [this period] could well be remembered as the one in which life on earth began to be completely transformed by the effects of mankind--the new Creator.[3]

Robert Sinsheimer sounds similar themes, wondering, too, whether we are not approaching the level of the gods. Sinsheimer takes the metaphor one more step, imagining humans as authors of their own divinity:

> For the first time in all time a living creature understands its origin and can undertake to design its future. . . . We are an historic innovation. We can be the agent of transition to a wholly new path of evolution. This is a cosmic event.[4]

But others wonder about the price we have paid. Citing bGH, the Beltsville hogs, GEHR crops and other technologies, global critics such as Martha Crouch, Wes Jackson, Kirkpatrick Sale, Michael W. Fox, Vandana Shiva, Jeremy Rifkin, John Fagan, Mae-Wan Ho, and Jack Kloppenburg tell a different story. While none of them has yet set out in a systematic way the global case against ag biotech, each has contributed important arguments to it.

Here is the case in brief. Ag biotech will not help us to pursue our best ideals of farming because it will not increase food security or equitable distribution of food; it will not help us to stabilize rural communities, become local to our geographical places, or pursue an environmentally friendly form of agriculture; it will spell disaster for women and children in developing countries; and it displays a technophilic hubris we should renounce. To oversimplify, the basic argument goes like this:

1. Ag biotech is an inseparable part of modern agriculture (MA).
2. Every inseparable part of MA vexes nature.
3. Therefore, ag biotech vexes nature.

The purpose of this chapter is to defend this argument.

1. Ag biotech is an inseparable part of modern agriculture.

What is modern agriculture? When Bacon encouraged efforts to gain control over nature at the turn of the 17th century, farmers around the world were barely able to grow enough crops to support themselves and their families. Today, at the turn of the 21st century, modern farmers in the developed world boast the most productive agriculture ever known.

Enthusiasts may overwork the aphorism, "Never before have so few fed so many," but the claim is true nonetheless. In 1850, a farmer in the United States could feed roughly 7 people. In 1990, according to the American Farm Bureau Federation, a farmer could feed 128 people: 94 in the United States and 34 in other countries.[5] Ag scientists have learned how nature works, and farmers have learned how to "bind her," producing unprecedented yields of corn, soybeans, wheat and oats.

MA means increased efficiency of production, and increased efficiency means increased food security, less manual labor, and more disposable income. These benefits are acquired through a technique called rationalizing production, which means growing only those crops suited to a particular climate and region, specializing in one or two crops, using synthetic fertilizers and pesticides to control weeds and pests, and increasing yields in order to reduce costs to the consumer.

The result is that farms in the United States no longer fit the popular image of mixed farms, on which mother raises children and chickens while daddy raises rotations of oats, beans, corn, wheat, hay, and pasture for his dairy cows, beef cows, pigs, horses, geese, and sheep. In general, farms in the US are economic firms specializing in one or two commodities, producing goods with synthetic chemical inputs for off-farm consumption on the national and international markets. Often, farms use natural resources without having to internalize the environmental costs, as when growers irrigate corn with free water and ranchers graze cows on public lands.

Why does ag biotech require MA? Richard Lyng, former Secretary of the US Department of Agriculture, provides the answer. Ag biotech, he writes, is a strategy to improve farmers' profits. Defending government-sponsored agricultural research in this area, Lyng noted that new technologies will result in more private sector jobs. New technology will "improve the quality of life by developing new uses and new markets for farm products, improving farm efficiency, and strengthening farmer profitability." Lyng claims that the basic impetus of all government-sponsored ag research "is not simply to increase production," but rather "to find answers to . . . challenges," adding that a "current challenge in agriculture is to remain competitive in the world market."[6]

Ag biotech requires the practices and institutions of MA because agriculture is competitive, competition requires innovation, innovation requires research, and research is expensive. Without the chance for significant returns on their investments, private companies cannot afford to invest in the basic research needed to identify and sequence genes and, whereas the potential long term returns on ag biotech research are huge, immediate returns are nearly negligible. Only groups with very deep pockets can afford to pursue ag biotech. These groups are transnational corporations and nation-states, countries with highly developed science infrastructures and tax bases.

Countries with subsistence forms of agriculture and limited capital cannot afford to do the research needed to develop the technology. As Crouch explains, ag biotech depends on the "large, complex industrial infrastructure(s)" of developed countries.

> Purified enzymes require rapid, refrigerated transport; information about genes is stored and manipulated in computer networks; chemicals and machines used in isolating DNA and maintaining constant temperatures for tissue growth rely on chemical companies, centralized and inexpensive energy sources, and efficient marketing.[7]

Government and industry officials who praise ag biotech as the next step in the march of progress agree with their critics on this point: ag biotech is married to MA. When proponents defend investment on the grounds that ag biotech will strengthen the competitive position of MA's high volume low cost producers, it comes as no surprise to learn that the first ag biotechnologies to have reached the market are bGH and herbicide resistant crops, technologies that favor larger over smaller farms. Nor is it surprising that ag biotech research is oriented to solve problems such as viral resistance to head smut in field corn and pseudorabies in hogs, agricultural problems not found outside MA's monocultural intensive animal confinement system. Ag biotech cannot be decoupled from MA because ag biotech is designed to solve MA's problems.

2. Modern agriculture vexes nature

While MA's benefits are obvious, its costs were largely hidden until 1962 when Rachel Carson's book, *Silent Spring* exposed the environmental consequences of synthetic chemical use in agriculture.[8] William Kittredge's family lived through the transformation from traditional to modern agriculture on a 7000 acre ranch in southeastern Oregon. His autobiographical account describes the changes MA brought to their modest hay and cattle operation:

> For so many years, through endless efforts, we had proceeded in good faith, and it turned out we had wrecked all we had not left untouched. The beloved migratory rafts of waterbirds, the green-headed mallards and the redheads and canvasbacks, the cinnamon teal and the great Canadian honkers, were mostly gone along with their swampland habitat. . . .

> We could not endure the boredom of our mechanical work, and couldn't hire anyone who cared enough to do it right. We baited the coyotes with 1080, and rodents destroyed our alfalfa; we sprayed weeds with 2-4-D Ethyl and Malathion, and Parathion for clover mite, and we shortened our own lives.
>
> In quite an actual way we had come to victory in the artistry of our playground warfare against all that was naturally alive in our native home. We had reinvented our valley according to the most persuasive ideal given us by our culture, and we ended with a landscape organized like a machine for growing crops and fattening cattle, a machine that creaked a little louder each year, a dreamland gone wrong.[9]

Agriculture is an organic whole, and all of us are implicated. When we change part of our food production system, we change part of ourselves. Commenting on Kittredge's passage, the novelist Jane Smiley insists that one take Kittredge's "we" personally,

> for whether we know it or not, as long as we eat, we are involved in agriculture, and through it, we are making our world, like Kittredge's valley, "a blank perfection of fields."[10]

2.1 MA and ag biotech vex ecosystems

MA vexes nature in many ways. Consider the, admittedly low-level, risk of widespread catastrophy resulting from the escape of a single virulent organism. If a genetically modified organism (GMO) were to escape into the environment and compete successfully with naturally evolved species, the results might be horrifying. Jackson compares the potential damage to the damage caused by chlorofluorocarbons (CFCs), introduced to serve useful purposes, including the cooling and preservation of food. Released into the atmosphere, however, CFCs attack the ozone layer and have produced a hole over Antarctica.[11] Jackson notes that it took the synthetic chemical industry less than a hundred years "before they were finally able to come up with a substance that would destroy the ozone." We might wonder with him how long it will take the ag biotech industry to engineer a product with similar global repercussions.

The risk of catastrophe from a single GMO is probably much lower than the risk of releasing CFCs, but the lesson of the CFC story is that we cannot foresee the magnitude of all of the risks. As Kristin Shrader-Frechette and Paul Thompson remind us, risk means different things to scientists and consumers. For scientists, risk is associated with a variable number and is based on calculated probabilities. For consumers, risk is more qualitative than quantitative, as seen in the difference in the language we use to distinguish "risky" from "safe" behavior.[12] If the successes of MA lead us to think that we have conquered nature and learned how to make our technological interventions "safe," we may be overstepping the boundaries of our knowledge. It would be prudent to remind ourselves of the awesome power of nature for, as Norman Maclean writes in his account of forest fire, "the terror of the universe has not yet fossilized and the universe has not run out of blowups."[13]

If a single catastrophic event from ag biotech is not highly likely, however, the longer-term accumulated risks of ag biotech's marriage with MA are worrisome. Here the risks are not from a single chemical or organism but from years of lower level, seemingly unrelated, events the synergetic effects of which may be massively destructive over time. The Environmental Release Committee of the Council for Responsible Genetics holds that "large scale releases of genetically engineered microorganisms into the environment pose risks that cannot be evaluated at this time with the current state of scientific knowledge."[14]

> Can we sustain MA into the future? Not in Crouch's opinion. Soil erosion and compaction by machinery is resulting in loss of substrate nutrients and structure; water is being used at rates that cannot be replenished; chemicals with various short- and long-term effects are being applied to the agro-ecosystem in large quantities; balances of both beneficial and harmful non-crop organisms (mycorrhyzae, pathogens, and so on) are being disrupted by monoculture methods, expansion into and interference with adjacent native ecosystems, use of chemicals, etc.; and nonrenewable fossil fuels are required to make fertilizers and to run machinery.[15]

There are numerous ways in which ag biotech will perpetuate this kind of food system and so prolong the trajectories in which MA is already vexing nature. MA's tractors run on fossil fuels; its cooling systems for food transportation and preservation require CFCs; and its feedlots emit methane, a gas implicated in the problems of global warming.

MA vexes trees. We have denuded some forty percent of the Earth's tropical closed rainforests.[16] Two millenia ago, our classical forebears cut down all the trees that once covered mountains in Greece; two centuries ago, our American forebears cut down virtually all of the trees in the northeastern US, and our remaining oldgrowth forests are under attack. Tropical rainforests are being razed with chain saws at the annual rate of an area the size of West Virginia in order to raise crops or food animals. The loss of tropical rainforests is troubling for many reasons, not the least of which is that they help to extract carbon dioxide from the air, and photosynethically to release oxygen and store carbon. As E. O. Wilson puts it, these forests

> cover only 7 percent of the Earth's land surface, [but] they contain more than half the species in the entire world biota. . . . [They] are being destroyed so rapidly that they will mostly disappear within the next century. . . [17] [perhaps by the year 2035], close to the date (2050) that the World Bank has estimated the human population will plateau at 11 billion people.[18]

Madagascar, "possesser of one of the most distinctive floras and faunas in the world, has already lost 93 percent of its forest cover," and the coastal forest of Brazil is 99 percent gone.

The motive in cutting down forests is to raise crops for subsistence, but Wilson observes that the forest soil is not well-suited to agriculture. When rainforests are cut and burned, the resulting ash and decomposing vegetation release a flush of nutrients adequate to support new herbaceous and shrubby growth for two to three years. Crops usually grow well at first, but soil fertility declines within three years, quickly reaching levels that are lower than those needed to support crops without artificial supplements.[19]

According to one observer, biotechnology will have its greatest impact on forests as fast-growing, high-yielding varieties of genefactured trees are raised to meet the demand for wood products.[20] Varieties may be designed to grow in tropical climates, providing new trees for replanting in cleared forests as an economic resource for landowners. How desirable is this technology? As Daniel Janzen opines, if biotechnologists develop economically valuable plants or trees that thrive best in cleared rainforest, "it is `goodbye, rainforest.'"[21] Ag biotech seems likely to prop up the practices and institutions of modern forestry, thereby contributing to the destruction of previously undomesticated ecosystems.

With rainforests being destroyed at the annual rate of between 17 and 50 million acres per year, some predict that the planet's average temperature will rise 3 to 9 degrees Fahrenheit by the year 2050.[22] A global warming trend would have profound consequences for farmers. In the United States, farmers in the southwestern states of Arizona, New Mexico, and Texas might be the biggest losers because they rely on expensive water supplies to raise crops in desert-like conditions; a rise in temperature would end most forms of agriculture there. The Corn Belt might also be hard hit by higher temperatures, sending the lucrative corn-growing industry further north, into the Dakotas, Minnesota, and Canada. Massive adjustments in international trade and corresponding political power would probably accompany global warming.

MA vexes land and what lies beneath it. It took geological processes millions of years to create the fossil fuel reserves that lay untouched until a century ago. Within the last century we have used up approximately eighty percent of all of the fossil-fuel oil reserves discovered to date in North America.

The two principal actors in the drama of oil consumption are the automobile and agriculture. Agriculture has not always been a major drain on our oil reserves for, as Lester Brown points out, at the beginning of the 20th century, "the world's farmers were almost entirely energy self-sufficient. The sun provided energy for crops to grow, livestock provided fertilizer and animal power [provided energy] for tillage." To produce a ton of grain at the beginning of this century required virtually no consumption of fossil fuels and added little pollution to the atmosphere. Today the situation is different: "On the average, the world's farmers [now] use the equivalent of more than a barrel of oil to produce a ton of grain. Each year it takes more."[23] Burning fossil fuels produces carbon dioxide, a pollutant. Our use of fossil fuels in agriculture is not diminishing: North America produces one-fifth of all the world's grain but, to capture that market, we have increased our use of fossil fuels six times since 1950 alone.[24] At current rates, we will exhaust known reserves by the end of the 21st century. MA spends the earth's capital at a rate greater than the earth can replenish it.

Paul Ehrlich is pessimistic about our capacity to heal the earth. His doubts arise from an analysis of the human exploitation of what Ehrlich calls "net primary production" (NPP). NPP is a measure of "the energy that green plants bind into organic molecules in the process of photosynthesis." All living organisms need organic molecules to survive. When one species begins to capture more than its share of NPP, other species are denied the resources they need. According to Ehrlich's estimates,

> The human share of the unreduced potential NPP reaches almost 40 percent. There is no way that the co-option by one species [out of a total of 1.4 million species] of almost two-fifths of the Earth's annual terrestrial food production could be considered reasonable, in the sense of maintaining the stability of life on this planet. . . . (If, as expected, we double our population by 2050, we will need to commandeer a total of 80 percent of terrestrial NPP,) a preposterous notion to ecologists who already see the deadly impacts of today's level of human activities.[25]

Ehrlich's image for what is happening is striking; "Earth's habitats are being nickeled and dimed to death . . ."

MA vexes water. Using center-pivot irrigation systems, modern farmers are pumping dry the Oglalla aquifer, a huge reservoir lying under Nebraska. Farmers on the Great Plains, according to Donald Worster, annually extract an amount of water from the aquifer that is "more than the entire [annual] flow of the Colorado River. That resource, left over from Pleistocene times, once the largest natural storage system of its kind anywhere, now has a life expectancy of about 40 years."[26] According to Jackson's calculations, producing one pound of feedlot beef in Nebraska and Colorado requires eight thousand pounds of fossil water spread over crops, and sucked up from aquifers "many, many times faster than the aquifers can be replenished."[27] In the western United States, agriculture accounts for 80 percent of water usage.[28] In southern California, the United States' most important agribusiness region, farmers are having an increasingly difficult time finding water to irrigate their desert crops. Salt residues in the soil resulting from current irrigation is also a major problem.[29] Across the United States, pollution from nitrogen fertilizer and pesticides has been found in areas relying on groundwater as the main source of drinking water.[30] Then there is the problem of soil and farm chemicals runoff into major estuaries, sources of fish for human consumption.[31]

MA vexes soils. Glaciers took hundreds of thousands of years to deposit the soils of Iowa that lay virtually untouched until a mere 150 years ago, and they left behind a huge checking account. In the mid-nineteenth century when my relations broke prairie sod, Iowa had on average some six feet of topsoil, some of the most fertile in the world. Now, after four generations of withdrawals from the glacial deposit, there are on average but three feet of topsoil left. Using conventional tillage systems to raise monocultures of corn, we have managed to squander half of a precious natural resource.[32] How?

Industry scientists set out to develop a genetically hybrid corn seed that would improve yield. They were successful, but at the price of sacrificing the corn's ability to reproduce itself. Modern varieties (MVs) must be bred by seed companies and then sold to farmers. The development of the MVs went hand in hand with the development of the seedcorn industry. Simultaneously, discoveries in inorganic chemistry led to the development of the fertilizer industry, meaning that farmers no longer had to pay attention to the fertility of their soil. Rather than rotating corn with legumes that would fix the nitrogen sapped from the soil by the corn, farmers could keep corn yields up by applying higher rates of anhydrous ammonia. The fertilizer stimulated the growth of weeds as well as corn, however, creating a rich environment for insects and pests. Consequently, farmers needed herbicides to kill the weeds and pesticides to kill the bugs. The pesticides wash down into the groundwater and soil is either compacted by the machinery running over it or eroded into rivers. Meanwhile, farmers are buying more and more petroleum-based products as the price of oil escalates.

In the Cornbelt, the monoculture method entails a loss by wind, rain, and sheet erosion of 20 tons per acre of soil, or 2.3 bushels of black dirt for every bushel of corn harvested.[33] Erosion of 5 tons per acre is considered an acceptable, because naturally replaceable, loss. It is difficult to put an economic figure on the 20 ton figure, but some agricultural economists have estimated it at $4 per ton.[34] Hans Jenny illustrates how extractive these farming methods are:

> Under average farming conditions, over one-third (35 percent) of the nitrogen and carbon content [of previously undisturbed American soils] had been eliminated in the first fifty years [of plow agriculture]. In a prairie soil in Missouri, the actual loss in humus amounted to thirty one tons per hectare.[35]

Estimates of the overall economic value of the loss of soil command attention. One estimate of losses in the United States put the figure at $7 billion from cropland soil running off and forming silt in "navigation waterways, water storage facilities, drainage ditches, and irrigation canals, and interference with water-based recreational opportunities."[36] According to a more recent estimate that includes costs to human health and infrastructure by David Pimentel in *Science*, the figure is $44 billion.[37]

If MA is using up fossil fuels, waters, and soils, it is also, in what is the most curious irony of the story, using up plants. By introducing domesticated varieties of crops, MA erodes plant germplasm diversity. Substituting a small handful of crops for human or animal consumption spells the end both of wild and native varieties. "In Sri Lanka," writes Robert Rhoades,

> where farmers grew some 2,000 traditional varieties of rice as recently as 1959, only five principal varieties are grown today. In India, which once had 30,000 varieties of rice, more than 75 percent of total production comes from fewer than ten varieties.[38]

How will we retain biodiversity in the face of the homogenizing forces of MA?

2.2 MA and ag biotech vex animals

As food animal production became industrialized and concentrated over the past five decades, the interests of agri-industry became dominated by fewer and fewer large corporations interested in short term profits. At the same time, consumer tastes became standardized while McDonald's hamburger stands went up in every town. The result was increasing pressure on agriculture to standardize its genetic stock.

Animal scientists responded by breeding increasingly specialized animals. In the dairy and swine industries the number of breeds has narrowed dramatically. Where once there were a dozen or more dairy breeds, today the Holstein dairy cow has virtually eclipsed the others. Where once hogs were as diverse as farmers and markets, today's standardized consumer preferences have dictated a correspondingly standardized swine gene pool. In the 1920s, a promotional booklet called the single-toed Mulefoot hog, "the most hardy, prolific, prepotent, early maturing, easy feeding . . . greatest money-maker of any breed."[39] By 1990, however, there was only one Mulefoot herd in existence.[40] One of the most popular breeds, the old-type Berkshire, standard in the United States before it was crossbred to the Poland China some thirty years ago, is hard to find today. In Great Britain, experts estimate that there are less than two hundred individuals of each of three traditional breeds: the Large Black, Red Wattle and Saddleback.

We have not retained diversity in farm animal breeds, and the reasons are not difficult to find. One reason is that producers rely on a decreasing number of influential companies for their stock. Another reason is that US milk pricing policy rewards high productivity with few rewards for high fat content. Public policies give comparative advantages to farmers able to purchase concentrated feeds more cheaply than farmers relying on grass forages.[41]

As argued in the previous chapter, industrial agriculture seems to conspire against animals. Before the gene era began, selective breeders had already produced experimental, domestic, and food animals unintentionally bred with characteristics that caused them lives of pain and suffering, and levels of intelligence below the levels of their ancestors. Genetic engineering increases the speed and power with which we can design such animals.

We have also previously noted the ways in which ag biotech entails suffering for domestic animals such as TAs and TFAs. MA vexes wild animals, too, for when we farm we convert diverse natural ecosystems such as grasslands, wetlands, and forests into more homogeneous agroecosystems: corn, wheat, and cotton fields. The expansion of agricultural lands over virtually every square mile of the globe's temperate regions has destroyed habitat necessary for species to survive. The result is that the number and diversity of wild animal habitat has declined dramatically with a consequent loss in the number and diversity of species. According to Wilson, the current rate of species loss is the largest it has been in the last 65 million years. He adds that,

> If present levels of forest removal continue, the stage will be set within a century for the inevitable loss of 12 percent of the 704 bird species in the Amazon basin and 15 percent of the 92,000 plant species in South and Central America."[42] These percentages are especially troubling in light of the fact that even a slight reduction in a species' numbers often results in a disproportionate loss of genetic variation.[43]

Loss of wild animal, bird, and fish species is not, of course, confined to rainforests in developing countries. Since 1800, for example, Dorset County, England, has lost 80 percent of its heathlands and 68 percent of its chalk downland. Correspondingly, the number and distribution of the common blue butterfly has fallen precipitously in the same areas; the numbers of silver spotted skipper butterfly have declined 66 percent.[44] In this regard, MA is different from traditional farming. One study found that on farms in England where hedgerows have remained in place, fields have remained smaller, chemical use has been minimal, and pastures have remained permanent or semi-permanent, the average number of mammalian species is 20. On such farms there are on average 37 bird species and 17 butterfly species. On modernized chemical farms, the respective averages drop to 6 mammalian species; 9 bird species; and 8 butterfly species.[45]

MA is not hospitable to many species that claimed our bioregions before us. In the continental US, passenger pigeons, bear, bison, and other species no longer have the wilderness acreage necessary to live without human management. Other species survive only in zoos or game parks, with the consequence that their gene pools become increasingly homogeneous and successive generations more and more susceptible to disease. Individuals in populations of wild species not in danger face a very uncertain future when subject to the management of humans, because our mismanagement often leads to suffering, malnutrition and death by starvation.[46] In Wilson's words, species extinctions present us with a "great natural catastrophe" on the order of the catastrophes that brought the Paleozoic and Mesozoic eras to a close.

Wilson estimates that the relative rate of species extinctions "with humans on the globe is 1,000 to 10,000 times as great as it was before humans."[47] Since the beginning of agriculture, domesticated animals and humans enjoyed a mutually beneficial relationship. Humans benefited from companionship, and animals benefited from humane care. But the relationship changed in this century when we intensified the pressure to select for desirable cosmetic and economic traits. As MA increasingly required a standardized product, breeders became unconcerned with the interests of the animals except insofar as those interests coincided with the breeder's interests. The breeders' interests were almost exclusively economic, with the result that wild sheep, capable of producing about 1 kg of thick rough wool each year as protective insulation have been made over into virtual wool machines, producing some 20 kg of fine downy wool for sweaters each year.

What happened to the animal in the process? Whereas sheep naturally shed almost all of their wool each spring during their seasonal moulting period, intensively bred animals have lost most of their biorhythms and do not moult with seasonal regularity; they must be shorn by humans.[48] Wild cattle that once produced a few hundred millilitres of milk each year have now been made over into virtual milk machines capable of producing 15,000 litres.[49] We now breed food animals that cannot perform the biological functions characteristic of their species, such as turkeys that cannot fly and cows that will not care for their calves. MA has gone so far in changing the genomes and phenotypes of our food animals that philosophers assert, somewhat grandiosely, that we have *created* these "artifacts," that the animals are more like machines than like wild animals.[50]

As happens with inbreeding among humans, narrowing gene pools often brings unintended results. When companion animals are back crossed for anatomical features consumers consider desirable, the animals often suffer problems, such as respiratory difficulties, anatomical abnormalities, or sensory deprivation. The dog has perhaps been treated worst of all as we have selected for traits that render some dogs virtually blind, lame, or incapable of breathing. We have bred dogs that seem to loathe themselves as much as they hate others.

The story does not end with breeding, however. Because the rationalization of agriculture requires low cost and high volume, we sought methods by which to house food animals in closer and closer confinement. Raising the number of animals per space increases the numbers a farmer can take to market. There are obvious limits. When animals are crowded together, living conditions may become so stressful that pigs bite off each others' tails whereas chickens resort to cannibalism and self-destructive pecking. Confined to small spaces, veal calves suffer muscle atrophy and anemia.

If the breeding principles of poultry biotech continue, what will the future of farming be like?

Picture yourself fifty years from now standing in the middle of a huge antiseptic warehouse staring at rows of tan colored objects that look something like footballs. Shiny stainless steel pipes descend from the ceiling and disappear into mouth-like orifices on top of each object. Black rubber tubes are attached by suction cups to the bottoms. The only attendant in the building tells you that the pipes bring water and rations to what he calls "the birds," while the rubber tubes carry excrement and urine to a sewer beneath the floor. Every twelve hours each bird drops a no cholesterol egg onto a conveyor belt.

"Regular as clockwork," he adds with a wink.

You are staring at thousands of living egg machines, transgenic animals genetically engineered to convert feed and water into eggs more efficiently than any of their evolutionary ancestors, layer hens. The science fiction objects I am asking you to imagine are biologically descended from the germplasm of many species unrelated in nature, including humans, turkeys, and today's chickens, so the worker is not speaking in mere metaphor when he calls the objects "birds." But unlike today's poultry varieties, which are only treated as machines, these brave new birds really seem to be more machine than animal. For, in coming up with the new birds, poultry scientists have not only selected for the trait of efficient conversion of feed into eggs; they have also selected for lack of responsiveness to the environment. The result is not a bird that is dumb or stupid, but an organism wholly lacking the ability to move or behave in dumb or stupid ways. Scientific research shows that the egg machine's complete lack of any externally observable behaviors is paralleled by its lack of physiological equipment necessary to support behaviorial activity. The brain of the bird is adept at controlling the digestive and reproductive tracts, but the areas of the brain required to receive and process sensory input and initiate muscular movement have been selected against, bred away. The new bird not only has no eyes, no ears, no nose, and no nerve endings in its skin; it has no ability to perceive or respond to any information it might receive if it had eyes, ears, or a nose.

The scene, inspired by a remark of Bernard Rollin's, is fantasy. To my knowledge, no poultry biotechnologists are aiming at an industry of unconscious bird-like egg machines.[51] But why not? Is there any reason to think that such birds are not the logical culmination of MA's breeding principles?

2.3 Ag biotech and MA vex humans

Which humans face potential harm from ag biotech and MA?

2.3.1 Family farmers in developed countries

Secretary Lyng argued that ag biotech will improve the quality of rural life by improving productivity and efficiency. He meant, presumably, that ag biotech will produce new jobs and raise personal income, capital accumulation, and entrepreneurial activity. But the picture of modern rural life reflects the reality of a few large specialized crop or dairy farms, suggesting a different prospect for the quality of life of many rural residents.

Quality of life is not measured only by financial indices. Other factors include availability and cost of health-care, life expectancy, infant mortality, disease incidence, work loss due to health reasons, level and type of education, school dropout rates, achievement scores, college attendance rates, participation in adult education and retraining, crime rate, incidence of alcoholism, drug use, domestic violence, suicide rate, proportion of population receiving public assistance, voter turnout, citizen involvement in government affairs, vitality of volunteer organizations. Many Americans move to or stay in rural areas not to improve household income but to pursue a lifestyle. For our agricultural officials to imply that economic development should be the primary goal of ag research is for them to overlook many of the features of farm and rural life that make it attractive.

When we consider urban and suburban sprawl and the demise of smaller farms over the last century, many will think initially that a Berry-like vision of a nation with more farms is deeply unrealistic. Nonetheless, we must ask ourselves about our moral principles and cultural vision, and whether our nation would be better off with lots of small farms than with a few large ones.

A classic study by the sociologist Walter Goldschmidt gives several reasons to prefer the populist vision. Goldschmidt studied two rural areas in California, Arvin and Denuba.[52] Arvin was surrounded by a few large corporately owned farm firms, Denuba by lots of traditional family farms. Goldschmidt found that Denuba had twice as many small businesses as Arvin; 60 percent more retail businesses; a higher level of per capita income; more self-employed people; more civic and voluntary organizations; more schools and more churches; and more citizen involvement in the schools and churches. By every standard of measurement Goldschmidt could think of, the family farm community had a higher quality of life than the large-scale agriculture community. Goldschmidt's work confirms the view that as farms became larger in California, the overall quality of life for residents declined.

In 1781, Thomas Jefferson wrote that "those who labour in the earth are the chosen people of God, if ever he had a chosen people."[53] In 1832, Andrew Jackson suggested that "The wealth and strength of a country are its population, and the best part of that population are cultivators of the soil. Independent farmers are everywhere the basis of society and the true friends of liberty."[54] In 1844, a letter to the editor of a journal wrote that

> "the farmer is the main support of human existence. He is the lifeblood of the body politic, in peace and war, . . . freedom, patriotism and virtue, after being driven from the degeneracy and corruption of the cities, will find their last resting place in the bosom of the agriculturalist."[55]

Popular mythology holds that farmers are harder workers, more honest, happier and live more stable lives than city folk. Is there any truth to it? A recent study by economists Renee Drury and Luther Tweeten, reviewing nineteen years of data from surveys conducted by the National Opinion Research Center at the University of Chicago, suggests that the answer may be affirmative. Farmers, they conclude,

> are less likely than others to agree that money is the most important thing in life next to health, . . . are among the least pessimistic, alienated, and fatalistic of all groups. . . . (C)ompared to the general population, the farm family is more stable and the typical farmer is more religious, politically more conservative, and happier and more satisfied with some aspects of life.[56]

Farmers are happier on average than city people, but not, according to Drury and Tweeten, happier than nonfarm or suburban residents. Nor does the study justify the claim that farmers have a higher overall quality of life than city dwellers. But farmers do appear to belong to an older paradigm of human culture and values that need not pass away.

Ag biotech also represents risks for women.[57] Bacon described the domination of nature by humans in gendered language, suggesting that humans are males who must harness a female companion. The image directs attention to the place of women in history. Women's labor has typically been undervalued in the US. From at least the nineteenth century on, farm chores have been divided by gender, with men tending cash crops and women tending crops for family consumption.

In the antebellum South, African American women and children worked as slaves on plantations. In the North, women and children contributed their labor without compensation. The marks of patriarchy and colonialism are evident in such systems because, as Carolyn Sachs points out, "although women's subsistence labor was economically essential for the survival of the farm, women's subistence work was undervalued because it was generally nonmarket activity."[58]

Women continue to work under the disadvantages of an imbalance of power. As Heidi Hartmann argues, a "set of social relations [with] a material base . . . [enabling men] to control women," is preserved in today's farm economy through job segregation by sex:

> Low wages keep women dependent on men because they encourage women to marry. Married women must perform domestic chores for their husbands. Men benefit, then from both higher wages and the domestic division of labor. This domestic division of labor, in turn, acts to weaken women's position in the labor market. Thus, the hierarchical domestic division of labor is perpetuated by the labor market, and vice versa.[59]

Agricultural research as a profession exhibits the same division of labor. A poll by Busch and Lacy shows that far fewer women go into agricultural research than do men, and the sexual division of labor tends to mirror the historical division of labor on farms. Men take up disciplines such as animal science and agronomy related to production; women take up research related to domestic activities: nutrition, textiles, and home economics.

Can biotechnology help us to redress the patriarchal stratification of ag research and production? Sachs' answer to a related question is not encouraging when its implications for ag biotech research are considered:

> As long as the hierarchical sexual division of labor in agricultural science keeps women in disciplines that focus on the consumer, the distance between production agriculture and nutrition will only widen.[60]

Most of the world's farmers are women. They do not participate in MA; they live on subsistence farms in developing countries.

2.3.2 Subsistence farmers in developing countries

Farmers in developing countries often live on the edge of food security, but they are more likely to be able to meet their family's food needs when they rely on local gardens, the plots of their own households and those of their neighbors. When a developing country begins to rely on world markets to meet its citizens' nutritional needs, however, the country exposes its people to risk by encouraging them to enlarge their operations. To expand, farmers must stop growing subsistence crops eaten locally, like beans and rice, and begin to grow high volume grain crops, or crops for export not meant for human consumption at all, such as animal feed or rubber. Crops once grown to sustain the indigenous population are replaced by money crops that reward successful farmers but may make it more difficult for the general populace to eat.

In an attempt to develop their own economies, Third World countries may rely on large infusions of capital from international lending agencies, such as the International Monetary Fund or the World Bank. Money makes it possible for the governments of developing countries to help smaller farmers to expand. Angus Wright attributes the switch to MA in developing countries to deliberate policies followed by development agencies and state governments.[61] As a few successful capital-intensive, modernized, farms get larger in the Third World, however, former landowning peasants are displaced from the fields of their ancestors. If they then elect to stay in their home rural areas, they may be forced to accept wages below the cost of living.[62] If they move to urban areas, they may be unable to find employment at all. In an article titled "Biotechnology Is Not Compatible with Sustainable Agriculture," Crouch concludes that "poor people are generally better off nutritionally when they are able to grow at least some of their own food," because only then are they not at the mercy of volatile markets.[63]

Studies seem to confirm Crouch's opinion. Summarizing the findings of Ferroni, 1980, Kathryn Dewey agrees that MA has had negative effects on poor families in Peru:

> Ferroni's central finding was that dietary adequacy was strongly positively related to the proportion of home-produced foods in the family calorie budget. Thus, families with greater independence from the market economy were nutritionally better off.[64]

Or consider Brazil where the underclass might be expected to gain from ag biotech. Brazil has between 5 and 7 million pre-teenage adolescents who live in poverty and eat by sifting through garbage.[65] Meanwhile, the annual inflation rate in Brazil is 800 percent. Biotechnology surely cannot solve all of these problems, nor is it realistic to expect it to do so. But before the green revolution, most of the 500 million acres of arable land in Campo Cerrado were pastureland. With the use of fertilizers, however, the land became very productive, so that by 1985 roughly 2 million metric tons of soybeans were being grown there.[66] Soybeans are grown for several reasons, including their agronomic value in fixing nitrogen in the soil and their economic value as export commodities.

As production of export soybeans grew, however, production of black beans for indigenous consumption was displaced. The result was a lack of the principal food stuff traditionally grown and consumed by the *campesinos*. Riots resulted.[67] The technology of the green revolution seems to have led to a concentration of land in the hands of a few large farms producing crops for export, displacing peasants from farms and apparently decreasing the availability of low cost food. The resultant socioeconomic problems are not simply the result of technical changes in agriculture; they result from a very complex interaction of domestic and international economic policy decisions, cultural attitudes, and historical trends. Without addressing these wider problems, we cannot lay the entire blame for Brazil's problems at the door of MA and MVs. But we may ask whether women and children on farms in Peru and Brazil today would have been better off had their countries pursued a different path in rural development.

There are other examples of MA vexing farmers. In Egypt, the modernization of agriculture has translated into more landless farmers.[68] In the Philippines, commercial interests are depleting the stock of fish on which indigenous people depend.[69] And in various places in Asia, forests "with multiple uses are turned into pulp plantations of eucalyptus which support very little life other than their own."[70]

How will ag biotech affect this picture? Ghana, Togo and Brazil all depend heavily on cocoa butter to bring in money which they use in turn to service their debt to the World Bank.[71] Recent advances in tissue culture, however, may permit major candy manufacturers in the developed world to produce a cocoa butter substitute far more cheaply than real cocoa can be produced in tropical regions of the world. The fear is that such a development would bring swift ruin to countries depending on cocoa as an export crop and already stressed by heavy debt loads. As Crouch argues, ag biotech turns farming into a business concerned primarily with profits, with potentially harmful effects on ecosystems and people:

> By turning everything it touches into commodities, biotechnology also has the effect of making products and processes that fit more easily into the global market. For example, seeds that used to be saved by the farmer now must be purchased every year. Genotypes that used to be specific to a slope, soil type, and rainfall amount in a particular valley are replaced with a genotype that will grow in a whole region. Markets that respond to short term increases in production replace subsistence or local markets that respond to the need for a secure food supply in unpredictable conditions.[72]

Crouch notes that successful subsistence farming still exists in many developing countries, including Mexico, Jamaica, India, Sierra Leone,

Kenya, and the Philippines.[73] But ag biotech threatens these proven modes of farming.

In 1986, the United States decided to subsidize substantially its rice exports, allowing its rice producers to undercut competitors. Without financial reserves to help its growers through 1986 and 1987, Thailand's farm families were devastated by the steep drop of world market prices for rice. As Don Reeves points out, the consequence of the US decision to dump government-held rice onto the world market meant that rice prices were cut in half.

> The US government made up the difference to its 19,000 rice growers. Thailand could not make up the difference to its 4 million farmers, most of whom grow rice.[74]

Sudden changes in the global economy are difficult for US farmers but they can be life-threatening for smaller countries. Consider another example, sugar.

> Over the past dozen years, half the US sugar market has been taken over by high fructose corn sweetener because of a combination of high sugar support prices and low corn prices. To protect its growers, the United States repeatedly has reduced its sugar import quotas, wiping out tens of thousands of sugar-worker jobs in the Philippines and the Caribbean.[75]

The action of the United States government is understandable. When it raised interest rates in 1979 to control inflation, the price of US agricultural goods on the world market went up dramatically. This cut into the level of goods we were able to export, and European nations, driven by the same spirit of competition, moved in to capture markets. Since 1979, the Reagan, Bush, and Clinton farm policies have been directed toward recovering those lost markets.

Larger export markets help, in a way, to support the US family farmer or, at least, the agricultural status quo in the US. But the effect of the forces of globalization on other exporting nations can be extreme. The physical health of a people can rapidly decline if their economy fails. Productive and steady employment may disappear, and the result can be a dispirited and resentful populace. As Lacy et al. point out, the forces of globalization and agricultural research seem to be conspiring yet again against small peasant farmers. Corporations are actively looking for ways to produce *in vitro* substitutes for rice, sugar, coffee, cocoa, rubber, cotton, and tea. If found, these high-tech goods will be massed produced and sold at prices below those of farmers in developing countries, further adding to the problems of nations "struggling to work their way out from under mountainous burdens of debt and handicaps of malnutrition and illiteracy."[76]

There is also the problem of transnational corporations (TNCs) in the affluent North profiting from germplasm taken, sometimes illegitimately, from the poor South. The Rural Advancement Foundation International (RAFI) and Calestous Juma have done much to bring this problem to light.[77] For several years, RAFI has sounded the alarm about bioprospecting and biopiracy conducted by corporations.

> Northern-based institutions seek access to tropical biodiversity for the primary purpose of developing profitable products. No matter how convincing the rhetoric, conservation and equity are secondary issues. Once indigenous peoples share information or genetic material they effectively lose control over those resources, regardless of whether or not they are compensated. if genetic material derived from plants, animals or microorganisms is eventually patented, access to this material can be legally restricted by monopoly patents. No matter what the circumstances, indigenous communities must have the right to say "no" to bio-pirates or legitimate bio-prospectors.
>
> Some people believe that current levels of technology will allow Northern-based institutions to undermine the importance of traditional medicine and respect for indigenous knowledge.[78]

The RAFI paper suggests in its headline that such arrangements reflect and perpetuate the "commodification of the sacred" and show disrespect for indigenous knowledge.

Ag biotech seems likely to devalue local knowledge; alienate indigenous peoples from their native cultures; widen the gap between haves and have-nots; separate those who generate knowledge from those who use it; divide those who produce food from those who consume it; and increase the differential in power between developed and developing countries. The present political and economic system on which ag biotech relies seems to engender unfair comparative economic advantages for well capitalized farmers in the North while disadvantaging less capitalized peasant farmers in the South.

2.3.3 Scientists and taxpayers

Transnational corporations (TNC) are calling on university scientists to help them answer basic questions in molecular biology. In 1984, for example, they gave grants of $120 million for university research in this area.[79] Of all funds spent on biotech at the US State Agricultural Experiment Stations in 1987, some twenty percent came from private companies. Compare this figure to the national average of university research monies that come from commercial interests, three to five percent, and you see how cozy the alliances have grown between commercial firms and universities.

However, not all universities engaged in biotech research will prosper from their involvement, because benefits are likely to be concentrated at a relatively small number of institutions. In 1987, thirty-three of the fifty US states were actively engaged in promotion of biotech research and development. Yet three states accounted for more than fifty percent of the $145 million invested that year.[80]

What kind of science is the well-funded university pursuing? Science that tries to solve problems by breaking problems down into simpler, more manageable, components and then trying to find technological fixes for these simplified problems. Ag biotech is reductionistic science based on the principle that component parts must be held constant while others are manipulated. The manipulations allow the scientist to determine functions by comparing the variable parts with the controls. Ag biotech, writes Crouch, must, by its very nature,

> be planned in advance in a linear step-wise series of procedures, which flows from the model of gene expression. The molecule carrying the genetic code, DNA, is transcribed into an intermediary, RNA, following the template in the DNA. RNA is then translated, also in a linear sequence, into a string of amino acids in a protein. In order to engineer a gene, the arrangement and sequence of elements in the DNA must be ascertained, and manipulated. Thus the investigator conceives of the project in a fairly precise, directed way. If the project cannot be designed in a sequence of well-characterized steps, the engineering project will not be feasible. . . . In concept, both the problem and the solution must be simple. Only one problem can be addressed at a time. . . .[81]

While the achievements of such science are undeniable, not all scientists are sanguine about its usefulness in solving world hunger and environmental degradation, because it unnecessarily confines attention to one problem at a time. For example, consider the problem of lysine deficiency in an area in Asia where children are malnourished. The problem comes largely from children lacking access to a variety of foods, and getting almost all of their protein from rice, which has very little of the essential amino acid.

Crouch writes that she received a letter from a graduate student suggesting that ag biotech research could "insert a gene into rice that codes for a protein high in lysine, thus balancing the protein."[82] The student added that the research could be done in an international non-profit laboratory, and seed could be given away to the poor. Is this not an example of ag biotech using its reductionistic science to solve an important issue?

Crouch responds that the narrow way in which the problem is approached subverts the legitimacy of alternative, holistic, solutions. Consider alternatives to genetic engineering for solving the lysine-deficiency problem. In areas where rice has been grown for thousands of years, the diet has traditionally been supplemented with legumes, which lack methionine but are sufficient in lysine. Thus dietary protein was made complete by complementation.

Perhaps a system-level research program could be used to solve the problem of lysine-deficiency. In this kind of science researchers would ask questions such as, What happened in the agricultural system in this part of Asia to disrupt the balance between rice and legume consumption? Were crop rotations abandoned because of a shift from subsistence to export farming? Did the loss of lysine in the diet result from rice monocultures displacing mixed farms of rice, legumes, and vegetable cropping patterns?

By examining the agricultural system as a whole, we may be able to relieve lysine deficiency in a traditional way by assisting farmers in returning to older ways of farming. A system-level approach would try to solve many problems at once. An increase in legume production in rice areas could improve soil fertility, disrupt pest and pathogen cycles, provide more employment for rural people, and solve the problem of lysine-deficiency.[83]

Ag biotech is not the answer to the problems of developing countries, and alternatives are available. The alternatives are better, because they offer holistic ways of resolving several problems at once whereas ag biotech can address but one problem at a time.[84]

2.3.4 Future generations

Given all of the premises defended so far, the problem concerning the risks of ag biotech to future generations may be stated succinctly. Assume that MA is a human and environmental failure; that ag biotech requires MA; and that ag biotech will exacerbate the worst features of MA. It follows that if we continue to follow the high technology monocultural path of MA we will so ruin the diversity, resilience and productivity of our agroecosystems that future generations will be unable to grow sufficient food.

What is the connection between environmental degradation, biodiversity, and the food needs of future generations? Garrison Wilkes, a professor of biology at the University of Massachussetts, makes the tie explicit. When a farmer changes from planting ten or twelve crop varieties in an area to one or two, the diversity of local land varieties is quickly lost. Wilkes writes that "at present rates of extinction, as many as 60,000 plant species--one-fourth of the world's total--may be lost or endangered within the next 50 years. Meanwhile there are more mouths to feed than ever." [85]

Robert Rhoades puts the problem of feeding future generations in historical context. When agriculture began ten thousand years ago, he writes, there were roughly four million people on earth.

> Today that many people are born every ten days. If the trend continues beyond the year 2000, we will have to grow as much food in the first two decades of the new century as was produced over the past 10,000 years.[86]

If the key to meet the monumental demand of future generations for food is wild plants then future generations may lack the resources they need to feed themselves: a diverse basis of plant and animal germplasm, adequate soils, and clean air and water.

3. Therefore, ag biotech vexes nature

I have been advancing the argument that ag biotech, being an inseparable part of MA, will vex nature. This argument is predicated on the consequences of ag biotech being of a certain sort. But there is another kind of argument against ag biotech altogether: That ag biotech is objectionable even if its consequences do not turn out to be objectionable. It is objectionable intrinsically, simply for the kind of activity it is.

4. Ag biotech is intrinsically objectionable

The Greeks referred to concerns about our moral and spiritual character as concerns about "arete," or concerns about our excellences, powers, and virtues. Aretaic concerns are concerns about the kind of people we are and are becoming. Does ag biotech threaten to form us into a kind of people we ought not to become? Does it bring us powers and ideals we ought not to desire? When we begin to tamper with an animal's genes, do we disrespect the animals' intrinsic value, trying to play God with another living being? Gene splicing techniques bring scientific powers we have not possessed heretofore, allowing us to mix and match species.

The first to articulate the intrinsic objection to biotechnology may have been Jeremy Rifkin in his book, *Algeny*. "Algeny" is Rifkin's term for the biotechnologists' form of modern alchemy, a kind of mystical science that transforms living things into things they are not. Biotech, based on the theory of evolution, reduces living beings to lifeless pieces of information. As Rifkin puts it, for the gene splicing age,

> Living things are no longer perceived as carrots and peas, foxes and hens, but as bundles of information. All living things are drained of their aliveness and turned into abstract messages. Life becomes a code to be deciphered. There is no longer any question of sacredness or inviolability. How could there be when there are no longer any recognizable boundaries to respect?[87]

Rifkin decries the biological— and spiritual— boundary crossing that gene-splicing admits. As algenists, scientists want to "help nature in its struggle to "perfect itself," trying to upgrade "existing organisms . . . with the intent of `perfecting' their performance."[88]

> For the algenist, species boundaries are just convenient labels for identifying a familiar biological condition or relationship, but are in no way regarded as impenetrable walls separating various plants and animals.[89]

When we begin to regard plants and animals as nothing more than bits of information we have lost the idea that these life forms are sacred. To engage in ag biotech is not only to cross species boundaries, but also to erode the foundations of our view that human beings are sacred, that we have a purpose or *telos* toward which we are oriented. Biotechnology participates in a worldview in which we ourselves are desacralized, turned with all living things from which we have evolved into automata, machine-like units. The Beltsville hog is just one particularly graphic illustration of how ag biotech enthusiasts regard other living creatures as mere bundles of information to be manipulated.

In Rifkin's view, the practice of algeny is not confined to scientific laboratories; it has extended itself into all corners of our lives, taking over our view of ourselves and our world. Whereas once we regarded nature as sacred and ourselves as its caretakers, now we see it as a profane machine with us as its engineers.

> In all of humanity's past experience, living things enjoyed a separate, unique, and identifiable place in the order of nature. There were always rabbits and robins, oaks and ostriches, and while human beings could tinker with the surface of each, they couldn't penetrate to the interior of any. Now, . . .

> the redesign of existing organisms and the engineering of wholly new ones mark a qualitative break with humanity's entire past relationship to the living world. . . ..[90]
> To design life by engaging in ag biotech is to commodify it. In ag biotech, everything has a price, and everything becomes a fit object for buying and selling.

Rifkin goes so far as to suggest that it offends God to cross plants with weeds when the two species cannot be crossed by natural means of reproduction.[91] Should we violate species boundaries set up by "natural law?" This question may appear extreme to some plant geneticists and breeders, but it deserves the careful attention of anyone genuinely interested in the future of agriculture.

If Rifkin is right, we have come to regard life as nothing more than "a base biological material, DNA, which can be extracted, manipulated, organized, combined, and programmed into an infinite number of combinations by a series of elaborate laboratory procedures."[92] The implications of this worldview are alarming, because it allows us and our children to think we can

> tear into everything around us, devouring our fellow creatures and the earth's treasures, all in the name of doing good, of ridding the world of evil. What we are really ridding the world of is its aliveness . . .[93]

Wes Jackson echoes some of Rifkin's themes and attributes our willingness to allow our agriculture to run the kinds of risks to animals, ecosystems, and humans enumerated above not to a conspiracy of industry and government nor to ag businesses simply wanting to sell their products and make money. He thinks the fault lies in our ready complicity with the experiment of modern agriculture, a belief system or paradigm that makes us think that ag chemicals are inevitable. What Jackson calls the Cartesian "knowledge-as-adequate world view" supplies the beliefs, values, and ideals of what he and Wendell Berry call "the modern industrial mind." Jackson, therefore, is less concerned about what he calls the Beltsville "hog monster,"

> than about the human monster, created by our culture, the monster who sees nothing wrong with creating such a hog. . . . The modern industrial mind is predicated on the Cartesian knowledge-as-adequate world view, [the view that humans can attain knowledge of whatever subject they desire and therefore adequately control whatever they desire,] and what has it produced? Acid rain, perhaps global warming, chemical contamination of the countryside we have no evolutionary experience with, Three Mile Island and

> Chernobyl. . . . When we spread atrazine all over Iowa and Illinois, we presume to *know*![94]

The ideology of MA, writes Jackson, is founded on three key assumptions:

> (a) Nature is to be subdued or ignored,
> (b) The purpose of agricultural research and farming is to increase production, and
> (c) Agriculture is to serve as an instrument for the advancement of industry.[95]

These assumptions destroy local agrarian ways of life.

> The policy of growing crops for cash for export instead of for local consumption may buy radios, but it will buy radios at the expense of soil erosion and chemical contamination of land and water. But that is to be considered progress [given these modern assumptions]. Progressive fundamentalism is as bad as religious fundamentalism, for fundamentalism takes over where thought leaves off.[96]

Jackson elaborates on the ideology that underlies this disrespectful attitude toward nature, tracing it to the French philosopher Rene Descartes and to the attitude that humans can know everything they need to know in order to control nature. He may just as well have traced it to Francis Bacon.

According to this view, biotechnology is an outward manifestation of an inner spiritual sickness at Michael W. Fox calls "technocracy." [97] Fox believes we need nothing short of a complete change in our worldview, a new paradigm that is "planetary and holistic" in which each of us realizes "the inherent wisdom of self-control in relation to the ecological whole (or unified field of being) and recognition of the intrinsic value of other beings."[98] Fox insists that his view is not anti-science and anti-technology, but fears nonetheless, that we have only two options.

> We can choose to engineer the life of the planet, creating a second nature in our image, or we can choose to participate with the rest of the living kingdom.[99]

5. What do global critics want?

Global critics want to replace MA with a paradigm whose beliefs, rituals, and ideals promote good farming. To change the practices and institutions of MA we need to change the beliefs and values that legitimate it. Critics of ag biotech think it will be no use as we try to change our attitudes.

What will be of use? New stories, borrowed in part from our oldest stories, to re-energize us with visions of the good life.

In his novel, *Remembering*, Wendell Berry describes young Andy Catlett, a reporter for *Scientific Farming* magazine and former farm boy assigned to write a feature article about Bill Meikelberger, the magazine's Premier Farmer of the year.[100] Meikelberger is a graduate of the College of Agriculture at Ohio State University, and owner of two thousand acres south of Columbus, Ohio, where farms average less than four hundred acres. Andy has heard of Meikelberger, and is excited at having such an important assignment. When he gets to the farm, he takes a quick look around and can see right off that Meikelberger's farm must have been

> the fulfillment of the dreams of his more progressive professors. On all the two thousand acres there was not a fence, not an animal, not a woodlot, not a tree, not a garden. The whole place was planted in corn, right up to the walls of the two or three unused barns that were still standing. Meikelberger owned a herd of machines. His grain bins covered acres. He had an office like a bank president's, . . . [and a house] with ten rooms and a garage, each room a page from *House Beautiful*, and it was deserted.

Having grown up on a much more modest farm, Andy had only seen pictures of farms like Meikelberger's. When the two of them go into the living room, Andy asks Meikelberger about his family. The farmer replies that he is all alone; his wife is in town at work, and his children all moved away. No need for them on the farm. When Andy questions him about his wife's work, Meikelberger grins and says "Every little bit helps."

Later, Andy sees him taking pills with his meal, and Meikelberger informs him that he has an ulcer. Leaving the farm, Andy reflects on Meikelberger's ambition, noting that this hero of modern agriculture,

> allowed nothing, simply nothing at all . . . to stand in his way: not a neighbor or a tree or even his own body. Meikelberger's ambition had made common cause with a technical power that proposed no limit to itself, that was, in fact, destroying Meikelberger, as it had already destroyed nearly all that was natural or human around him (pp. 73-76).

The problems of MA are not unconnected to humans. An agriculture that turns our valleys into blank perfections of fields cannot do so without at the same time impoverishing us. Berry's story, however also suggests an alternative road, a vision in which rural families, neighborhoods, and communities flourish.

In another scene, Andy is listening to his father, Wheeler Catlett, describe how Wheeler chose to farm. Years before Andy was born, Mr. Catlett had gone to Washington D.C. to attend law school and to work as a

Congressional aid. As Wheeler approached graduation, he was offered a job with a large packing house in Chicago. But did he really want to work as an attorney in Chicago in the middle of tons of concrete and thousands of pigeons?

Wheeler decided to return to his hometown to farm. Years later, when it comes time for Andy to decide on his career, Wheeler takes him out to the pasture. He tells him to look at the cattle, gathering in the walnut grove to drink.

> The cattle crowd in to the little stone basin, hardly bigger than a washtub, that has never been dry, even in the terrible drought of 1930; they drink in great slow swallows, their breath riffling the surface of the water, and then drift back out under the trees. Andy and Wheeler can hear the grass tearing as they graze (pp. 67-69).

Wheeler shows Andy the excellences of animal life, the virtues of the life of caring for animals, ideals best communicated not in arguments, facts, and figures, but in images, stories, and experience. A warm July night in the machine shed. A turn at the end of a field just planted. The cold air of a February high school basketball night in north-central Iowa. The sight of steam rising off a steer on a chilly May morning in Nathrop, Colorado.

Andy is supposed to travel to Pittsburgh that night, but as he leaves Meikelberger's award-winning farm, he enters another county, full of hills. He determines to take the back roads through wooded Amish country. The fields, he notices, are much smaller than Meikelberger's fields, and the farms more numerous. He meets one Isaac Troyer, who farms eighty acres with his wife, five children, and father and mother. Troyer invites Andy to dinner, but not before first inviting Andy to plow a few rows with Troyer's team.

As he drives back to Pittsburgh to write his story, Andy begins to wonder just which of the two farmers is the most progressive. He decides he will write about Isaac's "Premier Farm" rather than Meikelberger's. As he is driving something else suddenly dawns on him:

> Twenty-five families like Isaac Troyer's could have farmed and thrived--could have made a healthy, comely, independent community--on the two thousand acres where Bill Meikelberger lived virtually alone with his ulcer, the best friend that the bank and the farm machinery business and the fertilizer business and the oil companies and the chemical companies ever had.

Twenty-five and thriving on the ground now occupied by a single farm. Twenty five families, twenty five martin boxes, twenty five barns, twenty five orchards, twenty five stallions, fifty black mares, 75 children, and 375 guernsey cows, all flying and singing and eating and dying on ground now occupied by one man and his herd of machines.

This is what the global critics want.

Defenders of MA will reply that the idea of reviving small farms is attractive and quaint but highly unrealistic and politically naive. But our imaginations are powerful things, and stories can change the world. An alternative story that was at once powerful, true, and widely accepted could change our agricultural paradigm.[101] Such a story would not be widely accepted if it required us to sacrifice efficiency and productivity or if it required giving up food security. Nor should it, in my mind, require us to retreat from the quest for freer trade between the world's nations, or sophisticated large-scale communication and transportation technologies. Nor should it saddle farmers with acreage limitations, backward technologies, or a mentality that pits them against the conveniences and luxuries of contemporary society. Such a story must present an attractive vision of a new agricultural paradigm consisting of diverse small farms owned and operated by well-educated families connected up by computers and satellites in an international market system.

Berry points us somewhat in the direction of such a story by painting a fictional scene teeming with small farms. But his vision is not mere fiction, for Lancaster County in Pennsylvania boasts more than 5,000 farms, each farm averaging 84 acres. Seventy-five percent of those acres are under cultivation, in a rotation of corn, wheat, barley, oats, and hay. Eighty percent of the farms are owner-operated. In a recent year, each farm averaged $136,000 in gross annual sales.

Amish farms are not backward or inefficient. If efficiency is measured by the amount of energy consumed to the amount of calories produced, the Amish farm is more efficient than the Meikelberger farm with its heavy reliance on fossil fuel resources for energy. The farmers are relatively young, with an average age of 44, and they are, on the whole, quite content, because there is much good work to do on their farms, and because there is a future on their farms for their children.[102]

The Amish are not our only examples of good farming, and Berry recommends that we look for a diversity of approaches to good farming. He commends all farmers who practice non-degradatory agriculture. He salutes the 200-acre grass farms of Kentucky on which "less than 10 percent of the farm would be planted to crops that require disturbance and exposure of the soil," and slightly larger farms in the Cornbelt where "the cropping pattern is varied and complex."

There are several alternatives to modern agriculture. Our challenge is to tell their stories, and to devise public policies to help the stories continue.

6. Conclusion

The ag biotechnologies now on the market may be part of a huge and expensive technocratic food system, an undemocratic social and cultural nexus controlled by a scientific and engineering elite unconcerned with the interests of most of the world's plain citizens and farmers. In this essay, I have presented considerations that lean, in the style typical of academics who hedge every assertion with ten qualifications, in the direction of unconditional opposition. But now I lean toward declaring out right, that I globally oppose ag biotech.

At the beginning of this chapter, I quoted Benjamin Farrington's gloss on Francis Bacon's determination to vex nature. Science and technology have given us the skills to make nature "do what we want, not what she wants" and, since Bacon's death, agriculture, more than any other single field, has benefited from the richness of the metaphor of nature as a machine. With it, we have straightened rivers and irrigated hayfields, hybridized corn and tripled rice yields, invented engines, powered tractors, synthesized chemicals and killed pests.[103] The result has been an astonishing array of technologies making the lives of millions longer, better, and easier. But in the future, our deepest problems may not yield to solutions predicated on constantly finding new technological fixes.

Our challenge is to create a morally justifiable vision we can live by, a story based on a holistic, environmental, ethic. A story in which small family farms flourish; in which people seek the good of family and community before the good of individuals and corporations; in which children are taught to work hard, to honor those who have gone before them, and to be native to their places. We must compose together a story that encourages respect for animal life. We must learn that we are part of a larger pattern, a pattern, Jackson observes, "not of our making." [104] It is unclear at best whether any of the products of genetic engineering will help us to learn this lesson.

Notes

1. The quotation is a gloss of Benjamin Farrington's on section 98 of Francis Bacon's *Novum Organum*. See Farrington, *Francis Bacon: Philosopher of Industrial Science* (NY: Henry Schumann, 1949): 109-110. Here is Bacon's text: " . . . as in ordinary life every person's disposition . . . is most drawn out when they are disturbed--so the secrets of nature betray themselves more readily when tormented by art than when left to their own course." For Bacon, see *Advancement of Learning and Novum Organum* (NY: P.F. Collier and Son, 1900): 351. The last sentence in the passage from Bacon's "The Masculine Birth of Time," *The Works of Francis Bacon, Vol. III* (Philadelphia: A. Hart, 1853): 534.
2. Jeremy Rifkin, *Algeny* (New York: The Viking Press, 1983), pp. 50-51.
3. Andrew Scott, *The Creation of Life: Past, Future, Alien* (Oxford: Basil Blackwell, 1986): 190, 201.
4. Robert Sinsheimer, "The Prospect of Designed Genetic Change," *Engineering and Science Magazine*, California Institute of Technology, April 1969. Quoted in Leon Kass, *Toward a More Natural Science: Biology and Human Affairs* (NY: Free Press, 1985): 77. Kass adds in a footnote that "Dr. Sinsheimer has since had a chance of heart, and has become one of the advocates of caution and sobriety" (p. 351).
5. "Our comitment to providing safe and abundant foods," American Farm Bureau Federation flyer, 1990.
6. These statements by Lyng and Bentley are found in the USDA's "Yearbook" for 1986, a comprehensive review of the Department's scientific program. The book is widely circulated, and is regarded as a representative statement not only the USDA's research efforts, but of its spirit as well.
7. Martha Crouch, "Is Biotechnology Compatible with Sustainable Agriculture? No," *Ag Bioethics Forum* 4 (June 1992): 6.
8. Rachel Carson, *Silent Spring* (Boston: Houghton Mifflin, 1962).
9. William Kittredge, an essay. Quoted by Jane Smiley in her lecture, "A Thousand Acres: How Much is Enough?" Iowa Humanities Board, winter 1991, manuscript, p. 25.
10. Jane Smiley, "A Thousand Acres," p. 25.
11. G. E. Fogg and David Smith, *The Explorations of Antarctica: The Last Unspoilt Continent* (London: Cassell, 1990), pp. 179-180.
12. Cf. Paul Thompson on risk optimization in his *Food Biotechnology in Ethical Perspective* (London: Blackie, 1997), pp. 68-70, and Kristin Shrader-Frechette on rational risk evaluation, in her *Risk and Rationality: Philosophical Foundations for Populist Reforms* (Berkeley: University of California Press, 1991), pp. 66-77 and 169-217.
13. Norman Maclean, *Young Men and Fire* (Chicago: University of Chicago Press, 1992).
14 "Risk Posed by Large Scale Releases?" at http://nbiap.biochem.vt.edu/articles/apr9313.htm.
15. Crouch, "Very Structure," 151-158.
16. Wilson, "Current State," (1988), 10.
17. Wilson, p. 8.
18. Wilson, p. 10.
19. Wilson, p. 9.

20. T. M. Powledge, "Biotechnology Touches the Forest," *Bio/Technology* 2 (September 1984): 763-772; cited in Mark Sagoff, "Biotechnology and the Environment: What is at Risk?" *Agriculture and Human Values* 5 (Summer 1988): 28.
21. William Allen, "Penn Prof Views Biotechnology as a Potential Threat to Tropical Forests," *Genetic Engineering News* 7 (Nov-Dec 1987): 10. Quoted in Mark Sagoff, *AHV*, Summer 1988, p. 27.
22. Gail Wells, "Taking the Heat," *The Oregon Stater* (December 1988): 9-10.
23. "Official Sees Need to Ease Farms' Oil Use," *Des Moines Register*, 24 April 1988, quoting from *World Watch* magazine.
24. "Official sees need to ease farms' oil use," *Des Moines Register* 24 April 1988.
25. Paul Ehrlich, "The Loss of Diversity: Causes and Consequences," in E. O. Wilson, "The Current State of Biological Diversity," in Wilson, ed., *Biodiversity* (Washington, D.C.: National Academy Press, 1988), p. 23.
26. Don Worster, *Meeting the Expectations*, p. 61.
27. Wes Jackson, quoted in J. Tevere MacFadyen in "Wes Jackson: Taking on the Agricultural Establishment," *Country Journal*, July 1983, p. 74.
28. Pierre Crosson, "Sustainable food production: Interactions among natural resources, technology and institutions," *Food Policy* (May 1986): 143-156, citation at p. 150.
29. Robert A. Young, and Gerald L. Horner, "Irrigated Agriculture and Mineralized Water," in T. T. Phipps, P. R. Crosson, and K. A. Price, eds., *Agriculture and the Environment* (Washington, DC: Resources for the Future, 1986). Quoted in Reichelderfer, May 1989, p. 2.
30. Sandra S. Batie, "Agriculture as the Problem: New Agendas and New Opportunities," *Southern Journal of Agricultural Economics* 20 (1988): 1-12. Quoted in Reichelderfer, May 1989, p. 2.
31. Stephen R. Crutchfield, "Controlling Farm Pollution of Coastal Waters," USDA, Economic Research Service, *Agricultural Outlook*, AO-136, November 1987, pp. 24-25. Quoted in Reichelderfer, May 1989, p. 2.
32. In the Corn Belt, monoculture methods of raising corn entail a loss by wind, rain, and sheet erosion of 20 tons per acre of soil, or 2.3 bushels of black dirt for every bushel of corn harvested. Amory B. Lovins, L. Hunter Lovins, and Marty Bender, "Energy and Agriculture," in Jackson, *Meeting*, p. 81.
33. Lovins (1984), p. 81.
34. Cf. Steve Cain, "The Wisdom of Solomon," *Soybean Digest* (Mid-February 1991): 37.
35. Hans Jenny, "The Making and Unmaking of a Fertile Soil," in Jackson, *Meeting*, p. 49.
36. Katherine Reichelderfer, "Environmental Protection and Agricultural Support: Are Trade-offs Necessary?" National Center for Food and Agricultural Policy, Discussion Paper Series, FAP89-03, May, 1989, Resources for the Future, p. 2, citing Marc O. Ribaudo, "Reducing Soil Erosion: Offsite Benefits," USDA, Economic Research Service, Agricultural Economics Report No. 561, Washington, DC, September, 1986.
37. Pimentel, *Science* 267 (27 February 1995).
38. Robert E. Rhoades, "The World's Food Supply at Risk," *National Geographic* 178 (April 1991): 83.
39. Jerry Perkins, "A toehold on existence," *Des Moines Sunday Register*, December 3, 1995, Section J, p. 2.
40. Hans Peter Jorgensen, "An in situ model for preserving domestic animal diversity," The Institute for Agricultural Biodiversity's Preservation Breeder's Network," Luther College, Decorah, IA. Jorgensen cites Bixby, et al., *Taking Stock: The north American Livestock Census* (McDonald and Woodward, 1994).

41. "Preserving Herd's Diversity," *Des Moines Register* 26 April 1992.
42. Wilson, pp. 11-12.
43. Wilson, p. 11.
44. Goude, pp. 80-82.
45. Goude, p. 83.
46. We mismanage existent populations of wild animals by killing off predators. Sven Erik Jorgensen and William J. Mitsch, "Ecological Engineering Principles," in W. J. Mitsch and S. E. Jorgensen, *Ecological Engineering: An Introduction to Ecotechnology* (NY: John Wiley and Sons, 1989), p. 29.
47. Wilson, p. 13.
48. Goude, p. 67.
49. Andrew Goude, *The Human Impact: Man's role in Environmental Change* (Cambridge: MIT Press, 1981), p. 67.
50. Cf. claims to this effect by Paul Taylor and Baird Callicott.
51. See Bernard Rollin, *Between the Species* 2 (1986): 88-89.
52. Walter Goldschmidt, *As You Sow* (?) supply reference
53. Quoted in Gilbert Fite, p. 1.
54. Jackson, fourth annual message to Congress, quoted in Fite, p. 4.
55. Letter to the editor of the *Southern Cultivator*, in July 1844. Quoted in Fite, p. 4.
56. Renee Drury and Luther Tweeten, "Have Farmers Lost Their Uniqueness?" Anderson Report ESO 2237, Department of Agricultural Economics, The Ohio State University, 2120 Fyffe Rd., Columbus, OH 43210, 1995, p. i.
57. For an argument that feminists must oppose all forms of genetic engineering, see Linda Bullard, "Killing Us Softly: Toward a Feminist Analysis of Genetic Engineering," in Patricia Spallone and Deborah Steinberg, eds., *Made to Order: The Myth of Reproductive and Genetic Progress* (New York: Teachers College Press, 1987).
58. Carolyn E. Sachs, *The Invisible Farmers: Women in Agricultural Production* (Totowa: Rowman & Allanheld, 1983), pp. 4-5. Sachs cites Bengt Ankarloo, "Agriculture and Women's Work: Directions of Change in the West, 1700-1900," *Journal of Family History* 4 (1979): 111-120.
59. Heidi Hartmann, "Capitalism, Patriarchy and Job Segregation by Sex," *Signs* 1 (1976): 137-169. Quoted in Sachs, p. 72.
60. Sachs, *Invisible Farmers*, p. 63.
61. Angus Wright, *The Death of Ramon Gonzalez: The Modern Agricultural Dilemma* (Austin, TX: University of Texas Press, 1990).
62. See, for example, Wright (1990), and Susan George, *Food for Beginners* (New York: Writers and Readers Publishing, 1982).
63. Crouch, "Biotechnology Is Not Compatible with Sustainable Agriculture," *Journal of Agricultural and Environmental Ethics* 8 (1995): 98-111, citation at 102 and 103.
64. Kathryn G. Dewey, "Nutrition and Agricultural Change," in *Agroecology*, ed. by C. Ronald Caroll, John H. Vandermeeer and Peter Rosset (New York: McGraw Hill, 1990); citation to Ferroni at p. 464. Dewey is summarizing M. A. Ferroni, "The Urban Bias of Peruvian Food Policy: Consequences and Alternatives," Cornell University Ph.D. dissertation, 1980.
65. Raymond E. Crist, "Export Agriculture and the Expansion of Urban Slum Areas," *American Journal of Economics and Sociology* 48 (April 1989): 144.
66. P. H. Abelson, James W. Rowe, "A New Agricultural Frontier," *Science* 235 (20 March 1987). Cited in Crist, p. 145.
67. Abelson, Rowe. Cited in Crist, p. 145.

68. Edward Goldsmith, Nicholas Hildyard, Peter Bunyard, and Patrick McCully, "Whose Common Future? A Special Issue," *The Ecologist* 22 (July/August 1992): 144.
69. Crouch cites Kurien, 1991.
70. Crouch cites Lohmann, 1990.
71. Kloppenburg, "Defining the Challenges," p. 9.
72. Crouch, *Forum*, p. 6.
73 Crouch, "Very Structure," pp. 151, 156-157.
74. Don Reeves, "Linking US Agricultural Trade Policies and Third World Development," Bread for the World Background Paper No. 110, March 1989, p. 1.
75. Reeves, "Linking," p. 1.
76. Lacy et al., (1988): 6.
77. See Calestous Juma, *The Gene Hunters: Biotechnology and the Scramble for Seeds* (Princeton: Princeton University Press, 1989), especially pp. 149-169, and 228-237. RAFI, a research organization in Pittsboro, North Carolina, publishes the newsletter, "RAFI Communique."
78. For ideas used in this analysis, [RAFI writes], RAFI acknowledges the contribution of Jose Souza Silva, and his paper: Silva, Jose de Souza, "From Medicinal Plants to Natural Pharmaceuticals: The Commodification of Nature," presented at the Pan American Health Organization--InterAmerican Institute for Cooperation in agriculture, Symposium on Biodiversity, Biotechnology, and Sustainable Development, April 12-14, 1994, San Jose, Costa Rica.
79. Lacy, Lacy, and Busch, p. 7.
80. OTA, 1988, quoted in Lacy, Lacy, and Busch, p. 8.
81. Crouch, "Very Structure," p. 156.
82. Crouch, "Very Structure," p. 156.
83. Crouch, "Very Structure," pp. 151, 156-157.
84. Cf. Wendell Berry, "Whose Head Is the Farmer Using? Whose Head Is Using the Farmer?" in Jackson, *Meeting* (1984), p. 24.
85. Wilkes, quoted in Robert Rhoades, *National Geographic*, 1991, p. 83.
86. Robert E. Rhoades, "The World's Food Supply at Risk," *National Geographic* 178 (April 1991): 83.
87. Jeremy Rifkin, "The New Cosmic Mirror" in *Algeny*, reprinted as "A Heretic's View on the New Bioethics," in Laurence Behrens and Leonard J. Rosen, eds. *Writing and Reading Across the Curriculum* 5th ed. (NY: HarperCollins, 1994).
88. Rifkin, *Algeny*, 17.
89. Rifkin, p. 17.
90. Rifkin, p. 19.
91. Rifkin continues: " . . . By draining the aliveness out of things, we can pretend that our control and manipulation are of little consequence. . . . Our respect and reverence for nature diminishes as we gain greater control over it." *Algeny*, pp. 50-51.
92. Rifkin, p. 17.
93. Rifkin, p. 56.
94. Wes Jackson, "Our Vision for the Agricultural Sciences Need Not Include Biotechnology," *The Journal of Agricultural and Environmental Ethics* 4 (1991): 207-215, citation at pp. 207-8.
95. Jackson, "Vision," p. 208.
96. Jackson, p. 209.

97 Fox links the spiritual sickness of technocracy to an exploding world population, fearing that ag biotech will only make that problem worse. He notes that Molecular Genetics, Inc., an ag biotech company in Minnesota, recently won a $2 million grant from the US Army to genefacture a vaccine for Rift Valley fever virus, "a disease of cattle and humans prevalent in the Middle East and Africa." Fox is concerned not that the company will fail to produce the vaccine, or that it will produce it at too high a price, but rather that it will produce a cheap vaccine that is effective and useful. The result, he notes, will be an increase in the population levels of cattle and humans:

> This is a Catch 22 situation, since the increase in human population of a Rift Valley fever vaccine would create the need for an expanded livestock population to sustain the people. A vicious circle will develop without rigorous birth-control programs and the adoption of alternative agricultural and food habits.

Michael W. Fox, "On the Genetic Manipulation of Animals: A Response to Evelyn Pluhar," *Between the Species* 1 (1985): 52.

98. Fox, p. 52.

99. Rifkin, p. 252.

100. Wendell Berry, *Remembering* (San Francisco: North Point Press, 1988).

101. Cf. Berry's essay "Does Community Have a Value," in *Home Economics* (San Francisco: North Point Press, 1987) esp. pp. 189-192.

102. "Lancaster County: Farm Facts," brochure produced by Agricultural Committee, Lancaster Chamber of Commerce and Industry. Cf. Berry, "Whose Head?" pp. 25-26.

103. Cf. Leo Marx, *The Machine in the Garden: Technology and the Pastoral Ideal in America* (New York: Oxford University Press, 1964).

104. Jackson, "Our Vision," p. 214.

Chapter 5

Problems for the Case Against Ag Biotech, Part I: Intrinsic Objections

[I]t is important to see that the mere logical possibility of disaster is not sufficient to establish that the knowledge at issue ought not to be acquired. [For, were it so], we would quickly reach the absurd conclusion that [the research] both ought and ought not to be pursued. For just as it is logically possible that pursuing DNA research will lead to an unthinkable catastrophe, so it is logically possible that failing to pursue recombinant DNA research will lead to an unthinkable catastrophe.

- Stephen Stich[1]

I worked for many years constructing my version of the global case but, as I continued to try to strengthen it, I slowly began to lose confidence. My unease began with several personal experiences. One of our children had a common but annoying physical ailment, for which our pediatrician prescribed a very expensive nasal spray. When I inquired about its cost, the pharmacist informed me that it was a new, genetically engineered, product. The spray worked, and Karen and I never batted an eye.

Shortly thereafter, two of my diabetic friends independently disclosed that they were using a new, genetically modified, source of insulin. Each was satisfied with the product because it was cheaper and caused fewer side-effects, and was reportedly purer, than what they had previously used. I nodded approvingly.

How could I so readily accept medical biotechnologies while continuing to oppose ag biotechnologies?

I also met scientists who seemed to be counterexamples to the global case, researchers committed to using ag biotech for ends in which I believed. In 1991, I assumed the role of Coordinator of the Bioethics Program at my university. My responsibilities included the privilege of chairing a committee to plan a faculty development workshop intended to introduce discussions of ethical issues into life science courses. The workshop brought to campus various lecturers on ethical topics associated with ag biotech, including Martha Crouch and Wes Jackson. Both laid out formidable defenses of something like the global case. It seemed clear to those of us on the committee that these views demanded a response, and we looked for scientists to defend ag biotech. We first found Donald Duvick, retired chief plant breeder at Pioneer Hi-Bred, a private seed company in Iowa.[2]

Duvick delivered a paper in 1991 arguing that ag biotech would in the long run assist the cause of sustainable agriculture by providing diverse alternative crops genetically modified to suit the bioregions of different farmers.[3] His arguments planted seeds of doubt in my mind about two pillars of the global case: that ag biotech is environmentally unfriendly, and that it requires the paradigm of modern agriculture. Might we be able to decouple ag biotech from MA after all, and turn ag biotech to acceptable environmental and social goals?

Roger Beachy, a plant molecular biologist then at the Scripps Lab in La Jolla, California, also discussed his research with us. Beachy aims to help capital-poor subsistence farmers in developing countries by breeding new varieties of native crops capable of withstanding attacks by plant viruses. Beachy is working, for example, to bioengineer viral resistance into cassava. There is no world market for cassava, and yet cassava is an essential source of food for billions of people in the developing world.

Beachy pointed out that the techniques and implements of ag biotech become more affordable every day. He described a group of scientists in Vietnam using jerry-rigged equipment and tissue culture techniques to produce disease-free potatoes. And another group in Zimbabwe, who had constructed a bare-bones lab to genetically engineer virus-free fruit. Might it be possible to decouple ag biotech from the high-technology, monocultural, export-driven system of contemporary farming and integrate it into the local economies of less advantaged nations?

I continued to work at strengthening my global argument throughout the mid-1990s, but I was never sure that I had good answers for the questions raised by defenders of ag biotech. Personal experiences with medical biotech and two individual genetic engineers pursuing agricultural research goals in which I believed might not count for much for those approaching ethics with an abstract calculus. But, as I explained in the Introduction, first-hand experiences are significant, and offer pragmatic resources for solving problems.

Anecdotal experience, however, was not decisive in the end. Scientific and, alas, abstractly philosophical weaknesses in the global case were ultimately the factors leading me to change my mind.

Logically, my difficulties began with the observation by Stephen Stich that at least two uses can be made of the claim that biotechnology may have disastrous future consequences. One may use this claim to argue against the technology, or one may use it to argue for the technology. The problem can be put most clearly by considering the precautionary principle.

Many opponents of GM crops, and especially those in Europe, appeal to a particular philosophical principle to support their views. As formulated in the 1992 Rio Declaration on Environment and Development, the precautionary principle (*PP)* states that

> lack of full scientific certainty shall not be used as a reason for postponing cost-effective measures to prevent environmental degradation.

PP implies that we should not go forward with a new technology unless we are certain that it will be safe for humans and the environment. The principle is a clear expression of our natural risk-aversion and is so intuitively appealing that is has been codified into international law. Indeed, the European Union (EU) has invoked *PP* to justify its current moratorium on GM crops.[4]

The EU is correct that an implication of *PP* is to halt ag biotech.[5] But is this its only implication?

Suppose global warming intensifies and comes, as some now darkly predict, to interfere dramatically with food production and distribution. As we noted in ch. 2, massive dislocations in international trade and corresponding political power would follow global food shortages, affecting all regions and nations. In desperate attempts to feed themselves, billions would begin to pillage game animals, clear-cut forests to plant crops, and cultivate previously non-productive lands. Those with access to fertilizers and pesticides would begin to apply them at higher than recommended rates. The less fortunate would be forced to hunt animals of endangered species. Previously non-endangered species would be put at risk of extermination by marauding bands of humans. The human population would, as Michael W. Fox fears in ch. 4, launch a massive assault on what Leopold calls the land.

Perhaps not a likely scenario, but not entirely implausible, either. GM crops could help to prevent it, by providing hardier versions of traditional lines capable of growing in drought conditions, or in saline soils, or under unusual climactic stresses in previously temperate zones, or in zones in which we have no prior agronomic experience.

On the supposition that we might need the tools of genetic engineering to avert future episodes of crushing human attacks on the environment, *PP* requires that we go forward, full speed, with GM crops. Yes, we lack full scientific certainty that developing GM crops will prevent environmental degradation. True, we do not know what the full financial cost of GM research and development will be. But if GM technology helps to save the land, few will not deem that price cost-effective. So, according to the terms of *PP*, lack of full scientific certainty that GM crops will prevent environmental degradation shall not be used as a reason for postponing this potentially cost-effective measure.

Logical analysis shows that the precautionary principle commits us to each of the following propositions:

(i) We must not develop GM crops.

(ii) We must develop GM crops.

Yet (i) and (ii) are plainly contradictory, obliging us to perform two incompatible actions. The policy implications of the precautionary principle, therefore, are incoherent.

As a result of thinking about the relevance of Stich's argument to innovations in farming, I grew increasingly skeptical about objections to ag biotech based on unspecified claims about future disasters that ag biotech *might* cause. And I began to wonder whether I would have to change my mind about ag biotech the way I had changed my mind about animal rights.

What of family farms? What of the vision of a countryside teeming with barns, bats, birds and boys?

I continue to believe that ag biotech will vex small and medium sized farms. But this worry no longer carries the significance for me that it once did. An extended explanation of this point is in order.

New technologies are adopted because they use resources more efficiently. If the resource being used more efficiently is labor, then efficacious new technologies inevitably reduce the need for labor in the long run. This process happens in every industry, from the production of salt shakers to software.

Every new ag technology also harms some farmer or other because it eventually contributes to what economists call the rationalization of the industry. As efficiency of production goes up and more and more goods are produced more and more cheaply, the price the consumer must pay for the goods goes down. Consequently, fewer and fewer farmers can remain in the industry because they must have more and more land over which to spread their costs. Yes, ag biotech will almost certainly play a role in the demise of family farming because it increases efficiency of production. How could I, so committed to family farms, even consider accepting this consequence?

First, because it seems a fait accompli if the US has already lost its family farms. In the fifty years following 1940, before any ag biotech products had come onto the market Iowa, for example, had lost three-quarters of its farms.[6] Future historians will not place the blame for the loss of the family farm on ag biotech; they will place it on fertilizers, tractors, pesticides, international markets, and high yielding varieties, all of which came along before bGH and transgenic animals.

Second, because it may be impossible to do otherwise. Which new technology would favor all and only family farms? By definition, advances in technology bring comparative advantages for some and not others, with greatest advantages enjoyed by those who first use the technique. The principle holds true for all sizes of farmers. Just as first adopters of advances in MA enjoy comparative advantages over other MA farmers, so first adopters of advances in organic, sustainable, or permaculture farming enjoy comparative advantages over other organic, sustainable, or permaculture farmers. The dislocating effects of technical change are not confined to MA. Technical changes in every farm sector and farm size category work against someone, even if the change has been designed to assist a targeted class.

The point bears underscoring. Imagine a technology specifically intended to bring advantages to smaller over larger farmers, say, a small fuel-efficient tractor or a new easy-to-use, hand-applied, natural insecticide. To the extent that the technology succeeds, it will bring advantages to some small farmers. But not to all small farmers: only to those who adopt first.

It seems there is no middle ground here. Either we stop technological innovation altogether, or we accept the fact that it will inevitably displace some. As a result, we have two choices. Either we adopt the neo-Luddite line and oppose *all* ag technologies on the grounds that they will eventually drive some farmers out of business. Or, we bite the bullet, acknowledge that every new ag technology will inevitably harm many farmers, and set to work to devise cultural strategies to help displaced farmers find other lines of work. Not surprisingly, the second strategy is practiced by farm families throughout history, including the Amish, who know that, try as they might, they cannot place all of their children in farming.

Few of us are prepared categorically to oppose all technology, since that position quickly reduces to self-contradiction for anyone interested in using the telephone, fax, or email. We must accept not unregulated technological change, but some form of technological change nonetheless. As we do so, we ought to find ways to provide effective social mechanisms by which we can alert farmers to the inherent dangers of business in a technologically "progressive" world, mechanisms that can assist them in making adjustments and transitions. But it seems unfair to single out ag biotech for condemnation.

I have, alas, come to believe that we have already lost the family farm; that ag biotech played no role in its loss; and that no new technologies can be counted on to revive it.

Consider again the first point: While family farmers have continued to go out of business in the last two decades, the losses cannot be pinned on ag biotech. In my 1988 essay, I interpreted Kalter's data as predicting that bGH would drive fifteen to twenty-five percent of all dairy farmers out of business. But dairy farmers were driven out of the dairy business in comparable numbers by market and regulatory forces long before bGH was available for commercial use. Magrath and Tauer predicted in 1986 that as many as 5400 dairy farms in New York would fail in a three year period if dairy price supports were reduced or removed. The Magrath/Tauer prediction came true before 1990. During the relevant period, however, bGH was hung up in regulatory safety tests and not available for use until 1992. In sum, thousands of New York dairy farmers went out of business between 1986 and 1992. However, bGH, not being on the market, played no role in their decisions.

Is it fair then to criticize bGH, as I did in ch.1, on the grounds that it will lead to future injustices?

I have been making negative arguments against the claim that ag biotech will bankrupt family farmers. There are also positive arguments to be made on behalf of bGH. Let us return to the argument mentioned in ch. 1. If lower milk production costs lead in turn to lower food prices, lower food prices will benefit most those at the bottom of the ladder.

Luther Tweeten, an agricultural economist at Ohio State University, argues that the important point about bGH is not that the technology increases the amount of milk each cow produces but, rather, that it increases the efficiency with which each cow produces. Increased efficiency at the animal level leads to lower milk production costs at the farm level, and lower production costs on the farm translate in turn to lower milk prices at the supermarket level. Milk is a staple food, meaning that poor people must consume an amount of it roughly equal to that consumed by a rich person.

But are savings at the farm level actually passed on to consumers? Relying on comprehensive empirical data from a study by Kinnucan and Forker, Tweeten argues that decreases in farm dairy product prices are eventually enjoyed as savings by food consumers. To make the case, Tweeten reviews the literature on different dairy policy scenarios, adoption rates, likely consumer acceptance percentages, and bGH scale neutrality. He suggests that the data shows that lesser well-off American consumers benefit more from bGH than well-off Americans. According to Tweeten's calculations, American families grossing over $40,000 a year will save some $13 a year in milk and dairy product purchases if bGH is implemented. Poor Americans, those making less than $10,000 will save some $7 per year. On the face of it, this hardly seems to be an equitable result, with the rich receiving twice as great a savings from bGH as the poor.

However, the numbers must be adjusted for family size and marginal utility of income because rich and poor families typically differ in size and, while both kinds of families consume roughly the same amount of dairy products per person per year, the value of the respective dairy savings varies between rich and poor. For someone making over $40,000 per year, an extra $13 in the pocket means comparatively less than an extra $7 in the pocket for someone making less than $10,000. Because the demand for milk and dairy products is steady and inelastic, the bGH savings as a percentage of the rich family's income is estimated at 0.02, whereas for the poor family, it is 0.2. In other words, the poor family's saving of $7 per year is 10 times more important to them as the rich family's saving of $13. Tweeten concludes that "the net benefits from bGH would be distributed more equitably in relation to consumers' income than [any other] farm . . . technologies."

If bGH were not to be used, writes Tweeten, we would in effect be charging poor families a "milk tax" of $7 per year.[7] It seems difficult to think of a justification to ask the poor to pay a milk tax of $7 per year. To save family farms? But slightly higher milk prices will not be sufficient to save family farms.

Progressively lower milk prices, combined with progressively lower prices on other food items, are positive benefits to consumers, but especially to the least advantaged. Lower prices come from increased efficiency of production. In the US, yields of corn and soybeans tripled during the present century.[8] In 1900, Americans spent about forty percent of their income on food, whereas, in 2000, the figure has been cut to about fifteen percent. Increased efficiency of production has also enlarged the food choices of consumers in developed countries. In 1941, there were on average some fifteen hundred food items in grocery stores; today the figure is closer to fifteen thousand.[9] Taken as a whole, MA also brings benefits to those farmers with comparative advantages. They have lower input and labor costs overall, and must spend fewer hours in the field making passes over the crops. In 1910, the number of hours of labor required to produce a bushel of corn was fifty times as great as it is today. Consequently, farmers who are rapid adopters and good managers have more hours available to spend, say, with family members.

These positive arguments on behalf of ag biotech have been accompanied in my own experience by a renewed understanding of the realities of the farm I care most about. The Pippert farm continues to be owned and operated by a family, but it is no longer the family's sole source of income and it is now incorporated. The traditional family farmer rotated many different crops, integrated animals into the recycling of nutrients, received very little income from off-farm activities or government subsidies, produced most of the food consumed on the farm, and was not incorporated. The Pipperts no longer employ these strategies.

As my aunt and uncle have helped me to understand, their farm is not the farm celebrated in the children's song, "Old McDonald." Nor was it in 1988. Uncle Harold and Jason successfully raise corn and soybeans as cash crops in a two-year rotation, using the best available techniques of modern agriculture. Paying off college loans, they cannot afford to make a living raising a few chickens, hogs, and sheep. Now that Misty has died and her paddock and stall are empty, there are few animals on the farm other than the German shepherd, Kulo. Aunt Sandy, an accomplished teacher whose work in the community is widely respected, contributes substantial off-farm income to the household and has a wondrous garden. But she buys most of her groceries at the Hy-Vee.

I provide this information in the interest of a more accurate view of a typical Midwestern farm, and to explain my diminished confidence in the propriety of using tax monies to support commodity prices as a way to save the traditional mixed farm of decades past. Given a level playing field, some farmers will find innovative strategies to stay on the land. The Pipperts have remained in farming in part by renting and buying additional land, in part by diversifying their operation. Harold contracts with a local pudding producer to haul away milk left-over in the factory's pipes after a day of production. He incorporates the organic slurry into his soil, building up its fertility and contributing cash income to the farm operation. If it is wrong-headed to use taxpayers' dollars to support commodities (tax dollars that provide the largest benefits to the largest farmers), or to ban new technologies, it is not wrong-headed to expect that many Midwestern family-owned and family-operated mid-sized farms will find ways to survive on their own.

As my hope for mixed, smallish family farms diminished, my strongest objection to ag biotech lost its hold. At the same time, my worries about potential future environmental GMO catastrophes were outweighed by my belief that environmental damage might actually be more likely *without* GMOs. And the idea that ag biotech was an inseparable part of MA was undercut by the projects of scientists using genetic engineering for crops in developing countries.

I did not give up easily on the global case. Knowing that it has many facets, I determined to consider each on its own merits.

Fourteen intrinsic arguments

The global case against GMOs consists of two kinds of arguments, extrinsic and intrinsic.[10]

Extrinsic objections focus on the allegedly harmful consequences of GMOs, and argue that ag biotech should not be pursued because of its anticipated results. Briefly stated, the extrinsic objections go as follows. GMOs may have disastrous effects on animals, ecosystems, and humans. Possible harms to humans include perpetuation of social inequities in modern agriculture, decreased food security for women and children on subsistence farms in developing countries, a growing gap between well capitalized economies in the Northern hemisphere and less capitalized peasant economies in the South, risks to the food security of future generations, and the promotion of reductionistic and exploitative science. Potential harms to ecosystems include possible environmental catastrophe, inevitable narrowing of germplasm diversity, and irreversible loss or degradation of air, soils, and waters. Potential harms to animals include unjustified pain to individuals used in research and production.

Intrinsic objections to GMOs maintain that the process of making GMOs is objectionable in itself. This belief is defended in several ways, but almost all of the formulations are related to one central claim, the unnaturalness objection:

> *It is unnatural to genetically engineer plants, animals, and foods* (**UE**).

If **UE** is true, then scientists ought not to be engaged in bioengineering, however unfortunate may be the consequences of halting the technology.

Of the two sorts of arguments, intrinsic objections are the more powerful because if they are legitimate, then we should not develop GMOs, full stop. If society comes to accept **UE** as the conclusion of a sound argument, then much agricultural research must be terminated immediately and potentially significant benefits from the fledgling industry sacrificed. A great deal is at stake.

There are at least fourteen ways to defend **UE**.

(1) **To engage in ag biotech is to do what finite beings *cannot do*: transfer genes from one species to another.**
(2) **To engage in ag biotech is to *play God*.**
(3) **To engage in ag biotech is to *invent new technology*, an activity that should be reserved to God alone.**
(4) **To engage in ag biotech is to *invent world-changing technology*, an activity that should be reserved to God alone.**
(5) **To engage in ag biotech is to *arrogate historically unprecedented power* to ourselves.**
(6) **To engage in ag biotech is to exhibit *arrogance, hubris, and disaffection*.**

(7) To engage in ag biotech is unnatural because it is to *transfer the essence* **of one living being into another.**
(8) To engage in ag biotech is unnatural because it is to *change the telos*, **or end, of an individual.**
(9) To engage in ag biotech is illegitimately to *cross species boundaries*.
(10) To engage in ag biotech is unnatural because it is to *use nonsexual means* **to reproduce.**
(11) To engage in ag biotech is unnatural because it *causes harm* **to sentient beings.**
(12) To engage in ag biotech is to *commodify life*.
(13) To engage in ag biotech is to *disrespect life by patenting it*.
(14) To engage in ag biotech is unnatural because it *disrupts the integrity, beauty, and balance of creation*.

Consider each claim in turn.

(1) To engage in ag biotech is to do what finite beings *cannot do*: **transfer genes from one species to another.**

Were we to assert this claim at any time in history prior to 1981, we would be on firm ground, if we meant that we lacked the power to transfer genes across species lines via microinjection. But during the last two decades the ground has shifted, and the facts now give the lie to this premise. These days, scientists transfer genes from one species to another on an hourly basis. So **(1)** is a straightforward empirical claim. And it is false.

The unnaturalness objection is not usually intended as an empirical claim, however. Rather it is often formulated as a normative claim, as follows.

(2) To engage in ag biotech is to *play God*.

In a western theological framework, humans are creatures, subjects of the Lord of the Universe, and it would be impious for them to arrogate to themselves roles and powers appropriate only for the Creator. God created plants and we ought not to think that plants were put here for us to exploit. Shifting genes around between individuals and species is taking on a task not appropriate for us, subordinate beings. Therefore, to engage in bioengineering is to play God.

There are several problems with this argument. First, there are different interpretations of God. Absent the guidance of any specific religious tradition, it is logically possible that God could be a Being who wants to turn over to us all divine prerogatives; or explicitly wants to turn over to us at least the prerogative of engineering plants; or who doesn't care what we do. If God is any of these beings, then the argument fails because playing God in this instance is not a bad thing.

The argument seems to assume, however, that God is not like any of the gods just described. Assume that the orthodox Jewish and Christian view of God is correct, that God is the only personal, perfect, necessarily existing, all-loving, all-knowing, and all-powerful being. On this traditional western theistic view, finite humans should not aspire to infinite knowledge and power. To the extent that bioengineering is an attempt to control nature itself, the argument would go, bioengineering would be an unacceptable attempt to usurp God's dominion. So what's wrong with this argument? Simply that not all traditional Jews and Christians think that this God would rule out genetic engineering. Here the problem is the plurality of views, not between very different religious communities, but within a single, relatively homogeneous, religious community. Traditional theists disagree with each other about God's character and the scope of things God does not want humans doing. Consider Judaism. In the mystical traditions of the Kabbalah, God is understood as One who expects humans to be co-creators, technicians working with God to improve the world. At least one Jewish philosopher, Baruch Brody, has suggested that biotechnology may be a vehicle ordained by God for the perfection of nature.[11]

And why not? If humans are made in the divine image, and if God desires that we exercise the spark of divinity within us, then it should be no surprise that inquisitiveness in science is part of our nature. Creative impulses are not found only in the literary, musical, and plastic arts. They are part of molecular biology, cellular theory, and evolutionary genetics, too. It is unclear why the desire to investigate and manipulate the chemical bases of life should not be considered as much a manifestation of our god-like nature as the writing of poetry and the playing of sonatas should be. As a way of providing theological content for **UE,** then, argument **(2)** is unsatisfactory because ambiguous and contentious. There are two more theological interpretations of **UE**.

(3) To engage in ag biotech is to *invent new technology*, an activity that should be reserved to God alone.

Some of the literature attacking bioengineering takes a neo-Luddite line, suggesting that any new technology is suspect, that technology itself is the problem. These attacks typically are written on personal computers, printed on recycled paper and, often, disseminated through email across the internet. The difficulty here should be obvious. To oppose all technology is to deny our talents, presumably God-given. It seems counterintuitive theologically to commit oneself to a view that entails the conclusion that writing itself is unnatural. One need hardly note that the Scriptures themselves would not exist were it not for the technology of writing.

But perhaps we are selling this argument short. Here is a stronger version:

(4) To engage in ag biotech is to *invent world-changing technology*, **an activity that should be reserved to God alone.**

Let us consider **(4)** in conjunction with the next objection.

(5) To engage in ag biotech is to *arrogate historically unprecedented power* **to ourselves.**

The argument here is not the strong one, that biotech gives us divine power, but the more modest one, that it gives us a power we have not had previously. Given the astonishing practices of transgenesis, such as the movement of genes between fish and tomatoes, humans and hogs, one claim is obviously true: Ag biotech gives us power we have not previously possessed. But it would be counterintuitive to judge an action wrong simply because it has never been performed. On this view, it would have been wrong to do any of the following for the first time: prescribe a new herbal remedy for menstrual cramps; invent a new, more efficient, route for one's irrigation ditch; perform a Caesarean section; administer an anaesthetic; use a ballpoint pen. Much more is needed to call historically unprecedented actions morally wrong. What is needed is to know *to what extent* our new powers will transform society, whether we have witnessed prior transformations of this sort, and whether those transitions are morally acceptable.

We do not know how extensive the ag biotech revolution will be, but let us assume that it will be as dramatic as its greatest proponents assert. Have we ever witnessed comparable transitions? Probably. The change from hunting and gathering to agriculture was an astonishing transformation. Until ten thousand or so years ago (and in various locations yet today), people did not practice soil cultivation or animal domestication. Instead, they spent between 30 and 40 percent of their waking hours insuring that they would have enough food to eat and clothes to wear. They spent the balance of their time, several hours every day, dancing, playing drums, enjoying their children, going into trances, and telling stories.[12] Hunters and gatherers regarded the earth with religious devotion, and told sacred myths of a great Mother Goddess who blessed and cared for all animals; who required respectful treatment of her flora and fauna; and who blessed with food those who treated her with respect while withholding food from those who treated her with disrespect.

Eight or so thousand years ago, ancient brewers in Sumer and Babylon began making beer out of barley and hops while turning water into wine using grapes and vats. No one back then would have explained their actions this way, but they employed fermentation techniques on cereal grains, encouraging yeast cell organisms to swallow the grain and produce nutrients. Yeast grows by ingesting feedstuffs and giving off by-products, such as alcohol. Liquids shunned by the Women's Christian Temperance Union were not the only products of early bioprocessing, however. When the Hebraic tribes made their exodus from Egypt without leavened bread three thousand years ago they were temporarily doing without a fermented staple of their diet which their Egyptian hosts had already been making themselves for a thousand years.[13]

Around five thousand years ago, farmers in the Middle East raised grain using only human muscles, hoes, and digging sticks. When they began to capture, feed, and slaughter animals, they increased their food security. As they domesticated sheep, goats, and cattle, and worked to invent the plow, they figured out how to hitch cattle to the implement, were able to grow more grain than they could eat, and allowed the growth of cities. The kind of power that arose from the domestication of animals and the use of the plow can scarcely be overestimated. As William McNeill puts it, "That was how civilization arose--on the backs of the farmers."[14]

The historically unprecedented nature of the dawn of agricultural technology can be seen in the way the carrying capacity of the earth expanded in a relatively short period of time. Seventy five thousand years ago, world population stood at around five million. The population level remained nearly constant for more than 50,000 years and, about 10,000 years ago, population was still at five million. As the transition to agriculture provided huge increases in the amount of available food, however, population expanded dramatically. Within a period of four or five thousand years, population went from its plateau at five million to more than one hundred million. Along with this increase in the sheer number of humans on the globe came a rise in the amount and complexity of cultural activity. Writing, philosophy, music, the arts, politics, and architecture all got their start during this time. So what sort of power did people arrogate to themselves when they moved from hunting and gathering to agriculture? It is not hyperbole to answer: the power of civilization itself.

Some new technologies bring radically novel ways of perceiving and structuring the world. A new horse expands the horizons by tripling the distance one can cover in a day; a new plow doubles the amount of wheat I can produce in one year. Producing more wheat means finding ways to use or sell more of it. Selling surplus wheat means finding markets and seeking mechanisms to protect oneself from the vicissitudes of changing market prices. Changes in the kinds of animals and plants bred, eaten, and used by a people bring changes in the people, too. The traditional culture of the Plains Indians in the United States, for example, became much more mobile, aggressive, and well-fed when these native Americans began to capture and tame wild bands of horses drifting north out of what we now know as Mexico. When the Nez Perce nation began to breed horses selectively for agility and compactness, that nation not only introduced the Appaloosa to the world, but dramatically improved its own skills in hunting and warring.[15]

It is probably true that ag biotech brings us historically unprecedented powers. But this in itself is not an argument against ag biotech. On at least one prior occasion we have arrogated to ourselves historically unprecedented powers, and we are none the worse for it.

The objections stated in **(4)** and **(5)** are weak for two reasons. First, there is nothing intrinsically wrong with discovering and exercising new powers. Second, unless one thinks improved diet and food security are bad things and is prepared to object to agriculture itself, one cannot consistently object to ag biotech on the mere grounds that the transition introduces an unprecedented, world-changing, epoch.

(6) To engage in ag biotech is to exhibit arrogance, hubris, and disaffection.

It is certainly true that certain practices can dull one's sensibilities to pain and suffering. Many surgical students have an initial visceral reaction against cutting into human flesh, and the lengthy process of practicing surgery in residency is in part intended to help overcome the budding surgeon's aversion to the procedure. Some students in veterinary schools report a similar effect, that dissection in biology labs and junior surgery classes in vet school seem to make them less sensitive to animal suffering and pain. Thus it is possible that animal biotech as a practice might render some less sensitive to the well-being of research animals in particular and, perhaps, all animals in general.

But it hardly seems that the process of desensitization is necessarily a part of animal biotech, and there are certainly animal as well as plant biotechnologists who have not been rendered insensitive by their labwork.

Stephen Jay Gould does not disagree with Jeremy Rifkin's concern that we respect "the integrity of evolutionary lineages." And, as he points out,

> It would be a bleak world indeed that treated living things as no more than separable sequences of information, available for disarticulation and recombination in any order that pleased human whim.

And yet Gould is unconvinced that engaging in biotechnology will lead to a debased attitude toward life.

> I do not see why we should reject all of genetic engineering because its technology might, one day, permit such a perversion of decency in the hands of some latter-day Hitler. You may as well outlaw printing because the same machine that composes Shakespeare can also set *Mein Kampf*.[16]

Gould concludes that "the domino theory does not apply to all human achievements," therein identifying the problem with Rifkin's argument. **(6)** requires that we follow a slippery slope from relatively benign genetic engineering of plants and animals to starkly horrific genetic engineering of humans. But the slide down the slope is not inevitable, and can be blocked. There are counterexamples, including plant biologists using genetic engineering only to assist in the production of staple crops for the world's neediest people.

(7) To engage in ag biotech is unnatural because it is to transfer the essence of one living being into another.

This objection deserves extended discussion because it raises the following questions. When we transfer a gene from, say, a fish into a squash,

> are we conveying more than working base pairs? Are we conveying essence, somehow? Are we gaining not just genes in a new place, but also "ness"? -- fishness? Squashness? Human-ness? Will scientific explanations alone suffice to clarify this possibility? How do we account for our quizzical feelings that genes *convey,* something, somehow?[17]

The questions were originally put to me in a letter from Steven Burke inviting me to give a paper at a conference on ethics and genetic engineering. The questions are provocative because they suggest that the issue is moral rather than scientific. The question is not only, *Are* we transferring essences in genetic engineering? But also, If we reach the point where we are capable of transferring essences, *should* we do so?

First, we should not think that in transferring a gene we are transferring anything like an essence because in transferring chemical base pairs we may not change the organism. Imagine a petrie dish filled with bacteria. Into one of these bacterium we insert a single, inoperative, gene. It would seem odd to say that we have changed the bacteria, in any significant sense of the word changed, because the degree of change we have introduced could well be within the normal pattern of variation for this bacterial strain. Every strain of bacteria has a certain degree of variance, a family of typical chemicals that distinguishes the strain from other strains. But, all the while, a large number of other atypical chemicals come and go. Each bacterium regularly ingests foreign DNA sequences through insertions, mutations, and recombinations, or deletes existing DNA sequences. These changes often have no effect on the bacteria's functions. The bacteria is regularly changing, in a trivial sense, by taking on a variety of new chemicals in each generation. But to claim that the essence of the bacterium changes every time a stray chemical enters would make it impossible to identify any continuing essence for the bacterium.

The bacterial change just discussed provides a reason for believing that the action of inserting, say, a gene from a fruitfly encoding a protein responsible for the production of rosey color in the fruitfly's eyes into a bacterium is to change the bacterium only in the most trivial sense. It certainly is not to transfer the essence of the fruitfly into the bacterium. The moral: Transferring a fruit fly gene into a bacterium, even if it changes the bacteria's chemical structure, does not necessarily add any function or trait of the fruit fly to the bacteria. It certainly does not transfer `fruit flyness.'

But larger-scale changes to a bacterium's chromosomes certainly could change some of its functions. Might gross substitutions of genes change something's essence?

Suppose things have essences, that is, sets of properties that make them what they are: the intrinsic and indispensable conceptual characteristics of things. To transgenically transfer an essence via genetic engineering from organism *A* to organism *B* would be to move the essence of *A* into *B* where *A* and *B* were unrelated in nature, that is, could not recombine genes through natural means of reproduction.

There are at least two different ways of looking at nature. Essentialism thinks that the differences between, say, tomatoes and fish are grounded in two different sets of properties: the "essences" or "ideas" of the tomato and the fish. Essentialists are committed, as Elliot Sober explains to "there being some property which all and only the members of [a] species possess . . . some characteristic unique to and shared by all members of [the species] which explains why they are the way they are."[18] For essentialists, the essence of a tomato is fixed and unchanging. Applying this view to ethics, natural law essentialists would hold that it is immoral to tamper with the natural essences of things.

But the problem is that tomatoes are not fixed and unchanging. Just as there is tremendous variability within bacterial species, so there is tremendous variability within tomato species. Tomato genotypes vary within a single generation, and between generations. Modern evolutionary biology offers little hope for justifying the claim that there is a single essence identifying all tomatoes.

An example: Consider a single trait of the cherry tomato genotype, the height of the mature plant. We might ask, "What height is the `natural' height of this genotype?" If we are looking for the essence of the cherry tomato, the essential cherry tomato plant will surely have an ideal or `essential' height. But population genetics informs us that there is no answer to this question. You can take exactly the same cherry tomato genes and grow them under different environmental conditions and the height of the resulting identical plants will vary widely depending on the amount of sunlight, rain, and nutrients each receives. Conclusion: Not only are organisms constantly evolving and changing, but there is no "natural state" for any organism independent of the environment in which it lives. Rather, there is a wide range of phenotypes, which geneticists call the "norm of reaction," which identifies all of the individuals of a single species.

In sum, **(7)** fails for several reasons. We can transfer genes without transferring essences; it is impossible to identify the essence of a thing simply by describing its genome without describing its environment; and, in any case, it is unclear at best whether things have essences at all. Rather than worrying about changing something's essence, perhaps we ought to worry about interfering with their interests, causing them pain, harming them.

(8) To engage in ag biotech is unnatural because it is to *change the telos*, or end, of an individual.

Bernard Rollin has raised this matter forcefully in arguing that every animal has an end to which it is directed. He holds, with Aristotle, that each animal has a nature, "genetically based, physically and psychologically expressed," which determines "how they live in their environments."[19] Assuming that animals have a telos, genetic engineering might change an animal's telos. It may even be morally impermissible to change the animals' telos in some cases. But nothing follows from this argument to support the global case because there are a multitude of bioengineering transgenic animal projects that do not change the animal's telos. All such projects would be, all other things being equal, morally permissible, even under the telos assumption.

Consider a rather different version, consequentialist, version of this concern*: To engage in ag biotech is immoral if it changes the telos, or end, of an individual.* Rollin poses the problem this way:

> It is not inconceivable that as agriculture becomes more responsive to social pressure regarding confinement of animals, it will seize upon genetic engineering as a strategy for better fitting animals to their environments in order to reduce suffering. . . .[20]

Rollin comments that he does not consider such a strategy likely "in the foreseeable future," but at least one project has been going on for more than a decade. For fourteen years poultry breeders at Purdue University have aimed their research program at selecting birds that exhibit less stress in battery cages. And with success; the scientists have reduced mortality in cages from fifty percent to less than nine percent. The chickens have improved feather condition and no longer needed to be beak trimmed to avoid the birds killing each other.[21]

It may be that breeders have succeeded in changing the telos of the chickens, assuming that chickens have a telos. Is this a morally justifiable pool? Given only two choices, it would seem better *for the* birds to be incapable of being frustrated by conditions their ancestors would have found intolerable.

> If [Rollin writes] there were only two choices--either leave the animals as they are now, to live under conditions that do not meet their needs, or change their needs so they no longer suffer from the frustration of their fundamental urges--it seems clear that changing the animals is the lesser of the two evils.[22]

As Rollin is quick to point out, we do not have only two choices, and may leave the birds the way they are while changing the environments in which they are raised. Nonetheless, if we have only two choices, it would seem preferable, from the bird's perspective, to change its telos, if it has a telos.

In short, I fail to find any defensible interpretation of **(1)** through **(8)**, be it theological or secular, that provides good reasons to judge ag biotech so unnatural as to be intrinsically immoral.

There are other intrinsic objections to consider.

(9) To engage in ag biotech is illegitimately to ***cross species boundaries*****.**

This is an interesting argument because it captures our intuition that *something* is wrong in putting firefly genes into tobacco plants. But the problems here are both theological and scientific.

It is difficult to see how **(9)** could be defended on theological grounds. None of the scriptural writings of the western religions proscribe genetic engineering, of course, because genetic engineering was undreamt of at the time the holy books were written. Now, one might argue that such a proscription may be derived from Jewish or Christian traditions of scriptural interpretation. Talmudic laws against mixing "kinds," for example, might be taken to ground a general prohibition against inserting genes from "unclean" species into clean species. Here's one way the argument might go: For an observant Jew to do what scripture proscribes is morally wrong; Jewish oral and written law proscribe the mixing of kinds (e.g., eating milk and meat from the same plate; yoking donkeys and oxen together); bioengineering is the mixing of kinds; therefore, for a Jew to engage in bioengineering is morally wrong.

Let us assume that the basic principle is valid; it is wrong for observant Jews to mix kinds. The argument still fails to show that bioengineering is intrinsically objectionable in all of its forms for everyone. The argument prohibits Jews from engaging in certain *kinds* of biotechnical activity but not all; it would not prohibit, for example, the transferring of genes *within* a species, nor, apparently, the transfer of genes from one clean species to another clean species. To take a gene from a soybean plant and insert it into the chromosome of another soybean plant of the same variety is to engage in genetic engineering but it is not, apparently, to "illegitimately cross species boundaries" in the Orthodox Jewish view.

It is also worth pointing out that the Orthodox community seems to have accepted transgenesis in its food supply. Cheese is now routinely produced using a genetically engineered product, chymosin, and such cheese has been accepted as kosher, I am told, by Orthodox rabbis.[23]

There is another problem, which we can call the confessional problem. The confessional problem is the problem of trying to apply rules specific to a particular religious community, the confessing community, to the public at large. The confessional problem appears here in the following way. Some Talmudic laws are not meant to apply to non-Jews. Because it derives from the Jewish Oral tradition and not the Noahide Law, the law against mixing kinds constitutes a proscription to be observed by Jews, but not a law binding on those outside the Jewish community. In short, the law is a communal law of Jewish ritual not a universal law of ethics. It will, of course, be up to the Orthodox Jewish community to decide its attitude about genetic engineering. But the confessional problem will in any case block the ritual law against mixing kinds from serving as a basis for public policy. Consequently, the confessional problem will block the argument against mixing kinds from serving as support for the argument that bioengineering is intrinsically objectionable.

Consider another religious argument, that God established an internally connected self-organizing system with its own natural divisions. These divisions, including species boundaries, have a certain teleology, or end, and they are not to be conflated. Individual animals within one species can mate with others and so reproduce their kind, but God did not want individuals mating across the divisions. Humans, therefore, should not interfere with the natural directedness of the system by producing new species from two species unrelated in nature.

The problem with this argument was identified in ch. 2 (see discussion in "Unqualified Opposition"): species are not rigidly separated in the way implied, and species as a concept is context-specific. Members of different species give rise to members of a new species (mules). Differences between members of one species (dogs) can be greater than differences between members of different species (dogs and wolves).[24] Since nature gives ample evidence of generally fluid boundaries between species, proscribing the crossing of species borders seems ultimately to depend on religious assumptions. Without telling us what those assumptions are, however, we have no way of assessing whether we want to accept them.

Now, someone might respond to the mule counter example by arguing as follows. There are two kinds of species boundaries: one boundary between species close enough to be crossed through natural sexual reproduction, and another boundary distant enough not to be crossed through natural means. While the former boundaries, such as those between horses and donkeys, may be crossed, the latter should not be. Mules can be created by natural sexual means, but you cannot insert human genes into hogs by crossing a man with a sow.

The problem here is that species *transmogrify themselves* to produce new species. The story of evolution is the story of novel individuals; with each act of sexual reproduction, genes recombine to produce a unique phenotype. In addition to recombination, there is the additional possibility of gene mutation, migration, and incorporation, resulting in an individual which, if adaptively fit to its ecosystem niche, may be the founder of a species previously unknown in history. Considered as a process spanning billions of years, evolution presents countless instances of apparently unlike, unrelated, species which are in fact linked together. The problem with trying to distinguish two kinds of species boundaries, and then permitting the crossing of only those closely related, is that the argument assumes that species boundaries are distinct, rigid and unchanging. In fact, species are messy, plastic and mutable.

It is worth pointing out that even if the two kinds of boundaries argument worked, it still would lend no support to the view that bioengineering is intrinsically wrong because it would permit use of genetic engineering to transfer genes *within* a given species.

Yet another avenue of defense is available, however. One might claim that it is not the mixing of species that is objectionable, but rather the use of nonsexual means. This leads to:

(10) To engage in ag biotech is unnatural because it is to use nonsexual means to reproduce.

Some religious groups reject certain medical practices in the treatment of infertility on these grounds. For these groups, the following therapeutic regimes are forbidden: stimulation of ovaries to produce eggs, gamete intrafallopian transfer techniques, *in vitro* fertilization, and so on. If one rejects on theological grounds all nonsexual means of human reproduction, then one has a consistent position on which to oppose all plant and animal transgenesis.

There are two problems. First, much plant propagation is done asexually, from the child's scientific experiment of placing plant cuttings in glasses of water to the expert grafting of hybrid fruit trees. I am unaware of any ethical objections to these activities. Those who hold that asexual reproduction is morally wrong for humans typically do not object to it in the plant kingdom. Therefore, this objection would only apply to animal transgenesis; plant transgenesis would not be affected.

Second, the confessional problem raises its head if one tries to extend the objection to cover not only human but animal engineering. To ban animal bioengineering on the grounds that nonsexual means of reproduction is immoral outside the plant kingdom commits one to having to ban many commonly accepted infertility treatments for humans. Religious objections to the use of technology in assisting asexual intraspecific human reproduction may consistently be used to object to bioengineering as unnatural. But the conclusion will come at too high a price for all those not willing to concede medical procedures that have helped many infertile couples to conceive.

(11) To engage in ag biotech is unnatural because it causes harm to sentient beings.

Bioengineering may cause harm to sentient beings, such as humans and animals, if the plants it produces are toxic or mutagenic. But bioengineering does not inevitably cause harm. First, bioengineering benefits farmers and consumers who buy and sell its products. Second, plants are not sentient, so they cannot themselves feel pain as the result of being genetically engineered. As stated, therefore, argument **(11)** is false.

Suppose we changed the objection to: Bioengineering is unnatural because it sometimes causes harm to sentient beings. This claim may be true, but it does not provide an argument for objecting to all bioengineering, or even to bioengineering that causes harm. We often cause harm to sentient beings. Nurses cause children pain when they inoculate them, and professors cause students pain when they give them failing grades. Failing students may complain about their grades, and they may have a variety of objections to the grade, but it would be unusual for them to complain that it is unnatural for a professor to give such a grade.

I do not wish to be clever about this point, because some small scale farmers almost certainly will be made worse off by plant genetic engineering. If plant genetic engineering favors specialized farmers who can produce large quantities of low cost grains for export, then plant genetic engineering will harm peasants and mixed farmers by depressing the price of their products. This possibility rightly concerns us. My point is only that we do not clarify our intuitions here by adding to the concern about the suffering of farmers the idea that such suffering is unnatural.

(12) To engage in ag biotech is to commodify life.

The argument here is that genetic engineering treats life in a reductionistic manner, reducing living organisms to little more than machines. Along with Jeremy Rifkin, the United Methodist Church has made this argument, claiming that life is sacred and not to be treated as a good of commercial value only, to be bought and sold to the highest bidder.

Do those who object to ag biotech on these grounds apply the principle uniformly? Do they, for example, object to the commodification of cows, pigs, and chickens when farmers own and sell them on the market? One question we must ask is where we draw the line. If one accepts commercial trafficking in food animals, then it is hard to see why those animals should not be further commodified by genetic engineering. If one accepts the commodification of vaccines, to be bought and sold to treat disease, then one has committed oneself to accepting the commodification of thousands of species of animals, namely, all those at the microorganismic level. Why should it be unnatural to treat DNA the way individual food animals, plants, and microorganisms are treated?

"Life" is an ambiguous term covering a multitude of uses. It is wrong to commodify individual human lives, to buy and sell people. But is it wrong to sell organic parts of people? Perhaps we should avoid trafficking in scarce human organs and blood, and we ought to find a rational way to allocate livers, kidneys, corneas and such. But would it be wrong everywhere for anyone to traffic in human parts? Consider a mother contemplating selling her hair to make toupees. Would it be wrong for her to do so? Let us assume that she is very poor, lives in a developing country, and faces the prospect of her ten year old daughter entering the sex trade in order to provide food. In such a situation, it seems far better morally to sell pieces of one's hair than to force another to sell their body. We have good reasons to protect individual humans from being commodified. But there is much room to argue that, given careful legal fencing and scrupulous protection of the innocent, *parts* of humans might very well be fit objects to be treated the way we commonly treat corn, squash, eggplant, pigs, cows, and chickens. It is not irrational to regard pieces of DNA more on the order of pieces of cheese or hair, *parts* of human bodies, than to see them as individual lives, *persons* worthy of moral standing.[25]

For these reasons, I have come to believe that while **(12)** may be true, it probably is not a major concern. But there is a related issue:

(13) To engage in ag biotech is to disrespect life by patenting it.

There are two distinct arguments here.

(13.1) To patent the products of plant ag biotech is to devalue nonhuman life.

This argument has little to recommend it. The system of patent protection was extended to living matter in 1873 when the U.S. government awarded a patent to Louis Pasteur on a disease-free yeast he had manufactured. In 1930, patent protection was extended to asexually reproduced plants in the Plant Patent Act. In 1980, the Supreme Court ruled in *Chakrabarty* v. *Diamond* that a "man-made" microorganism, namely, a bacterium engineered to breakdown crude oil, was patentable. This last decision seemed to many to reverse the view held until Chakrabarty, that living matter was not itself patentable. Plants were considered patentable only because of the explicit act of Congress in 1930.[26]

Whether we are morally justified in patenting novel plant species seems to me no more debatable in the 1990s than whether we are justified in killing weeds. When a group objects to the patenting of the plant products of ag biotech because doing so devalues life, we must ask why this is so. Unless the objector also objects to the Plant Patent Act of 1930, the principle is being applied unfairly. To approve the patenting of hybrid corn seeds but not the patenting of transgenic soybean seeds is to be guilty of inconsistency. **(13.1)** proves too much.

(13.2) To patent the products of animal ag biotech is to devalue nonhuman life.

The patenting question does not seem compelling in the area of plant life, but it is a different matter when we cross into the animal kingdom. Sentient beings who can experience pain and emotion have interests which may well be thwarted if they are the product of two species unrelated in nature. In 1980, living things other than plants became eligible for patenting. The first animal patent was granted in April 1988 to Harvard University for the so-called oncomouse, a mouse genefactured for susceptibility to cancer. To accept the patenting of plants while rejecting the patenting of animals seems to be a plausible position, but it too has problems. To know that we are devaluing animal life by patenting it we must know how much value nonhuman animal life has. But animal life seems to have little value beyond its economic value for most people in developed countries where millions of animals are slaughtered daily for food. It is difficult to see how the mere act of patenting an animal devalues its life any more than the currently accepted practices of owning, artificially breeding, confining, and killing food animals at a young age, or caging research animals to conduct experiments upon them. Unless one takes a strong animal rights position in which it is morally wrong even to own animals, **(13.2)** assumes too much.

I conclude that we have not yet found a good reason to believe either of the variants of **(13).**

(14) To engage in ag biotech is unnatural because it disrupts the integrity, beauty, and balance of creation.

There are two ways to understand this claim.

(14.1) To engage in ag biotech is impermissible insofar it will have the consequence of disrupting the integrity, beauty, and balance of creation.

The more we learn about ecology, the more we understand how little we know and how much less we can control as massive and complex a system as a biome. It has taken nature millions of years to evolve the diverse species currently in existence, and species lines seem to have a coherence of their own. Since ecosystems exist in such a delicate balance, for us to think that we can manage them in a way that will preserve their ancient wisdom and beauty may be hubris. At the Vancouver Assembly of the World Council of Churches in 1983, the Council claimed that creation has *integrity*, which Jay McDaniel later would define as "the value of all creatures in themselves, for one another, and for God, and their interconnectedness in a diverse whole that has unique value for God."[27] This value is neither one created by humans nor one that humans ought to try to control. But doesn't bioengineering presume that we can control the future direction of evolution?

This objection has much to commend it, but I think it is not usually understood as an intrinsic objection. To argue that bioengineering will disrupt the integrity of creation is an objection to the possible *effects* of bioengineering. Interpreted as **(14.1)**, argument **(14)** is not a variant of the unnatural line of argumentation but a variant of the extrinsic objection to bioengineering's potentially adverse environmental consequences. Note, then, that argument **(14.1)** does not lend support to **UE**.

But there is a second way to understand **(14).**

(14.2) To engage in ag biotech is to *vex nature*, to disrespect its intrinsic value; therefore, ag biotech is impermissible.

Here we have an argument based on the metaphor found in Farrington's gloss on Bacon at the beginning of Chapter 4, and from which I have taken my title for this book, that ag biotech *vexes nature*. But what would it mean to "vex nature?"

To vex a person is to trouble, distress, annoy, irritate, or disturb them. I vex my third grade son Drew when I tease him about unknown girlfriends, deny him his request for his allowance, or punish him for straying into the street. To vex someone may be morally objectionable, and then again, it may not. Everything seems to turn on the circumstances. I am probably in the wrong when I tease him about girls, but in the right when I instruct him about the dangers of speeding cars. To vex persons is not necessarily to fail to respect them.

To vex an animal is to trouble, distress, annoy, irritate or disturb it. Morally permissible? Again, it depends. I am morally in the wrong when I vex Charlie, the wild horse at Honey Rock Camp, by hitting him in the head with a stick. But I may well be morally in the right to put a hackamore gently on his head. Both actions distress him, but one of the actions does not necessarily to fail to respect him. To vex an animal is not necessarily to act immorally toward it.

As the previous examples suggest, we can make sense of the idea of vexing individual sentient beings. To vex is to interfere with the individual's preferences, their goals. But can we make sense of the idea of vexing nature, which seems, after all, not to be an individual so much as a group of individuals, animate and inanimate? We may reply positively if we can show that nature has the kind of things individuals have: preferences and goals. Can we show this?

Many ecologists and environmental philosophers think so, and their affirmative answer should appeal intuitively to all who love flora and fauna. The major themes of an ethic in which nature is so construed were first sounded in contemporary America by the Midwestern conservationist Aldo Leopold in 1933.[28] Forty years later a professional philosopher named Richard Sylvan, his last name at the time was Routley, made an attack on agriculture the cornerstone of his environmental ethic, beginning his seminal 1973 essay, "Do We Need a New, an Environmental, Ethic?" with these words:

> It is increasingly said that . . . Western civilization . . . stands in need of a new ethic . . . setting out people's relations to the natural environment, in Leopold's words "an ethic dealing with man's relation to land and to the animals and plants which grow upon it."[29] It is not of course that old and prevailing ethics do not deal with man's relation to nature; they do, . . . man is free to deal with nature as he pleases . . . (M)en do not feel morally ashamed if they interfere with a wilderness, if they maltreat the land, extract from it whatever it will yield, and then move on. . . . Under what we shall call *an environmental ethic* such traditionally permissible conduct would be accounted morally wrong, and the farmer subject to proper moral criticism.[30]

In 1974, there was no recognizable philosophical school of environmental ethics but, twenty years after Sylvan's essay, a diverse array of extensively discussed positions has developed. Ecofeminists have an ethic based on the conviction that the historical subjugation of nature by humans is inextricably connected with the subjugation of women by men. For ecofeminists, we cannot address the issue of environmental abuse by modern agriculturalists without also addressing the issue of the abuse of women by modern patriarchalists.[31]

Biocentrists have an ethic based on the conviction that *all* living individuals, plant as well as animal, are owed respect and that one cannot exclude individuals from the circle of moral standing just because they are not sentient.[32] For biocentrists, individual human life is not necessarily superior to animal or plant life. For them, the received anthropocentric ethic is based on unjustifiable moral principles because it does not grant moral standing to living things outside the circle of individual *homo sapiens*.

Ecofeminism and biocentrism deserve consideration. However, I will focus my attention on a third group, containing direct descendants of Leopold and Sylvan, because it is arguably the most influential. *Ecocentrists* have developed the land ethic into an important philosophical theory that tries to bring considered judgments, moral principles, and background scientific theories into reflective equilibrium around the core notion that humans should not be free to deal with nature as they please. Rather, we should recognize intrinsic value in conscious and nonconscious wild living beings and nonliving natural objects such as rocks and soil.

John Rodman, for example, believes that "thistles, oak trees, and wombats, as well as rain forests and chaparral communities" have intrinsic value. Forests, he asserts, "have their own characteristic structures and potentialities to unfold," and thus ought to have a *prima facie* right not to be interfered with.[33] If Rodman is correct, then the rights of forests must be considered along with the rights of humans. The old anthropocentric principle must be revised, so that the intrinsic value of the environment can be weighed equally with the intrinsic value of human beings.

Ecocentrism holds that humans have historically conceived of themselves as the rulers and endpoints of nature and consequently have all but ruined it.[34] We must change our self image from that of rulers or even stewards and caretakers to that of beneficiaries or, as Leopold put it, "plain citizens" of the biotic community. We must no longer treat soil, water, and wildlife as valuable only for the instrumental use we can make of them because nonhuman life is intrinsically valuable. The way to effect this change is to shift our attention away from individuals to biotic communities. The new locus of intrinsic value must be biological wholes, including ecosystems such as deserts, prairies, and pine forests, and natural entities and processes such as the hydrologic cycle.[35]

The basic principle of Leopold's Land Ethic is that "a thing is right when it tends to preserve the integrity, stability, and beauty of the biotic community. It is wrong when it tends otherwise."[36] For Leopold, moral guidance comes first from nature, not from norms guarding the rights of rational agents. Ethics as a discipline, therefore, must be refashioned, based on scientific knowledge about how nature invents and preserves itself.[37] J. Baird Callicott is ecocentrism's ablest proponent. He describes the essential features of the theory here:[38]

> Its conceptual elements are a Copernican cosmology, a Darwinian protosociobiological natural history of ethics, Darwinian ties of kinship among all forms of life on Earth, and an Eltonian model of the structure of biocenoses all overlaid on a Humean-Smithian moral psychology. Its logic is that natural selection has endowed human beings with an affective moral response to perceived bonds of kinship and community membership and identity; that today the natural environment, the land, is represented as a community, the biotic community; and that, therefore, an environmental or land ethic is both possible . . . and necessary, since human beings have collectively acquired the power to destroy the integrity, diversity, and stability of the environing and supporting economy of nature.[39]

To represent the natural environment as a "community, the biotic community," is to think of ecosystems as "collective organisms," a metaphor Callicott borrows from Leopold, who in turn borrowed it from the organismic ecology of Frederick Clements, a plant successionist in the early part of this century. The model has been developed further by Eugene Odum, who stresses the interdependence of all plants and animals within each biotic community, describing ecosystems as integrated, stable systems existing in, or in the process of attaining the state of, mature equilibrium.

Organismic ecologists believe the stability of the mature ecosystem state is proportional to the diversity of species, and that the greater the number of species, the greater the stability; the fewer the species, the less stable the system. As Ned Hettinger and Bill Throop describe it,

> These states may involve some change, such as fluctuations in the populations of predators and prey, but such changes are regular and predictable (as in the cycling of predator and prey according to the Lotka-Volterra equations). Disturbance and change are thought to be atypical. When such a system is disturbed, it will gradually return to its mature state, a state characterized by a balance or harmony between...elements.[40]

According to ecocentrists, the scientific theory just described contains criteria for the moral evaluation of human interventions into ecosystems. An action is morally right for the land when it promotes the health of the ecosystem, when it protects or adds to the diversity of the species in the system, or when the action in some other way protects or adds either to the integrity and stability of the whole or toward the system's progress toward its goal. Because ecosystems have an endpoint, a good or welfare such as stability, then what is morally good or bad for the ecosystem, and what actions are morally right or wrong for us to take toward it, follow from understanding each ecosystem's mature equilibrium state. In general it will be morally right for us to engage in actions that will leave intact or enhance the diversity of flora and fauna present in a biotic community and morally wrong to engage in actions that will undermine the stability and integrity of the ecosystem's progress toward its mature equilibrium state. It will be morally good for us to leave ecosystems alone, and morally bad for us to disrupt them.

Because it is the overall equilibrium of the system that matters, and not the individual rights of specific organisms within the system, it follows that individuals exist for the sake of the whole, not the whole for the sake of individuals. Therefore, if the number of individuals of species *q* becomes so large that it overruns and displaces the individuals of species *r*, then the morally required act may be to kill individuals of species *q* so as to maintain stability and biotic diversity. The emphasis in ecocentrism on community and holism clearly distinguish it from traditional individualistic and anthropocentric ethics.

The ecocentrist's sense that farming vexes nature makes sense in light of this background scientific theory. Agriculture inevitably disrupts the equilibrium state of any ecosystem because it requires actions that decrease the number and diversity of species. Farmers systematically kill indigenous plants that compete with their crops, and exterminate animal species that endanger their food animals. But need ecocentrists disapprove of *all* farming? Might there not be some forms of farming that would be ecocentrically correct? To answer this question we must return to Sylvan's essay.

Sylvan gives three examples of conduct of which he disapproves. The first is a man who drives the Australian dingo to extinction in order to prevent the animals from interfering with Australian farm operations. The second is a fisher who kills the last remaining blue whale for private profit. The third is a farmer who, Sylvan writes, borrowing the words from Leopold, "clears the woods off a 75% slope, turns his cows into the clearing, and dumps its rainfall, rocks, and soil into the community creek."[41] Sylvan's three examples contain two farmers, both of whom he disapproves.

Sylvan's last example, the one taken from Leopold, is also picked up by Callicott. In an essay called "Animal Liberation: A Triangular Affair," Callicott does not search for reasons that the farmer might have committed as dubious an act as clear cutting a steep hillside, nor entertain possible excusing conditions such as the possibility that the farmer in an emergency needs the pasture to provide essential food for his family. Rather, the ecocentrist accuses the farmer of being "morally wanton" because of the effect of his action on the good of the whole.

"Wanton" means senseless, unprovoked, recklessly or arrogantly ignoring justice and decency. To call someone morally wanton is not to praise them. The farmer is reckless, according to Callicott, because the farmer who turns the

> dairy cows out to pasture in a woodlot situated on a steep slope overlooking a trout stream (for the [mere] sake of the shady comfort and dietary variety of the cattle) [commits an action] with ruinous impact upon the floral and wildlife community native to the woods, the fish and benthic organisms of the stream, and the microbic life and the physiochemical structure of the soil itself.[42]

Many would agree with Callicott that the farmer, excusing conditions aside, has done something wrong. But defenders of farmers might suggest a very different reason for this conclusion, basing their opinon on the principle that clearcutting a steep slope is just bad farming. If you clear cut, you will no doubt have a "ruinous impact upon the floral and wildlife community native to the woods," but there will also be disastrous consequences for your farm. Loss of soil, erosion damage to the slope, pollution of the stream--all of these actions will reduce the productivity, not to mention the beauty, of your place.

Notice how different the agriculturalist's principle is from the ecocentrist's. Ecocentrists hold that clearing the slope is wrong in and of itself because the woodlot ecosystem is morally considerable in itself. To harm the ecosystem, even if the harm consists only of harm to "the microbic life and the physiochemical structure of the soil itself," is to do direct harm to an intrinsically valuable thing. Farmers therefore are "morally wanton" not because they occasionally engage in actions that endanger the land's usefulness to future generations of humans and animals, but rather because they habitually and regularly engage in actions that contravene duties they have directly to the land itself regardless of any indirect deleterious consequences of those actions. Tillage inevitably destroys the integrity, stability, and beauty of the microbic life and physiochemical structure of the soil.

It is important to state explicitly the logical conclusion of the ecocentrists' position. It is not just bad or stupid farmers who are morally wanton. To farm at all is to act senselessly, recklessly, and arrogantly, even if you farm in the Amish, perennial polyculture, or traditional Tohono O'odham desert, way.[43] For even sustainable agriculture, inherently an anthropocentric notion, is a practice aimed at circumbscribing the extent to which we exploit the soil. According to the logic of ecocentrism, farming of whatever sort is wanton behavior.

The ecocentrist's science and values are not, of course, the only viable candidates for our allegiance, and we have other values we want to protect. The received ethical view holds that people have a right not to go hungry when food can be made available to them, and it is a widespread intuition that land has instrumental value and ought to be put into production to help keep people from going hungry. The moral paradigm that attributes basic moral rights to individuals is pervasive in the West, and the majority of individuals in democratic societies seem to share the intuition that the right to be fed when it is possible to provide you with food is one of of these basic rights. How, then, should defenders of agriculture who accept the received view reply to ecocentrists?

One might try to reduce the ecocentrists' position to absurdity by showing, for example, that it leads to unacceptable consequences. If ecocentrism were true, no one could plant a garden, much less half a section. If ecocentrism were true, no one would be able to survive. But a position that leads to that consequence must be absurd.

I do not think the *reductio* strategy will work, however. Ecocentrists can simply bite the bullet and admit that their view does lead to something like this conclusion. Self-described "bioregionalist neo-Luddites" such as Martha Crouch point out that humans apparently lived as hunters and gatherers for thousands of years before agriculture got started, and Callicott seems to think the world would be a better place, because a more wild and diverse place, if humans returned to our former feeding habits. We may reply: But the earth cannot support its present population level if we return to hunting and gathering. Again, the ecocentrist may affirm the conclusion: The number of humans on the planet *is* too high, and we ought to try to reduce it. Ecocentrists need not commit themselves to a reprehensible program of genocide and mass starvation; they can argue that we ought to embark on an aggressive program of birth control and rational planning so as to reduce the human population without killing anyone.

I do not think the *reductio* will work because I do not believe the ecocentrist position has counterintuitive results. It is perfectly plausible to think of hunting and gathering as an acceptable way of life, a way of life in which humans can flourish, and which, were it widely practiced, would lead to a much lower danger of species loss. Were there far fewer of us, we would be able to do a better job of sharing the earth with other species.

A stronger objection is that ecocentrism leads not to genocide, but to occasional sanctioning of murder.[44] Ecocentrism apparently requires us to sacrifice the life of an individual human if so doing is the only way to preserve the last remaining example of an endangered nonhuman species. A popular presentation of the ecocentric ideal makes this consequence explicit. In his novel *Desert Solitaire*, Edward Abbey, author of *The Monkey Wrench Gang* and noted defender of the deserts of the Southwest, laments the oppressive presence of humans in the United States' southwest Arches National Monument park. He opines " . . . I have personal convictions to uphold. Ideals, you might say. I prefer not to kill animals. I'm a humanist; I'd rather kill a *man* than a snake."[45] If confronted with the tragic choice, ecocentrists are required by their theory to kill innocent people rather than endangered animals.

For his part, the early Callicott, who once characterized the present number of humans in the world as "a global disaster...," accepted this criticism. While he later revised his opinion, Callicott once wrote that the true measure of an environmental ethic is the extent of its misanthropy, adding that Abbey "may not be simply depraved."[46] Callicott reasoned that Abbey was probably interested in "dramatically making the point" that in the imagined case, the choice between a human and a snake, "would be moot." Callicott went on to assert that his

> biospheric perspective does not exempt *Homo Sapiens* from moral evaluation in relation to the well-being of the community of nature taken as a whole. . . . As omnivores, the population of human beings should, perhaps, be roughly twice that of bears, allowing for differences of size. A global population of more than four billion persons and showing no signs of an orderly decline . . . is at present a global disaster (the more per capita prosperity, indeed, the more disastrous it appears) for the biotic community. . . . The extent of misanthropy in modern environmentalism thus may be taken as a measure of the degree to which it is biocentric. [47]

Are the problems presented in ch. 4 so severe that our only recourse is to become misanthropic ecocentrists? Or have ecocentrists overreacted? Tom Regan has famously charged Abbey and Callicott with being environmental fascists.[48] But, again, this line of attack begs the question of whether the foundations of our morality and civilization are the right ones. The premise of ecocentric ethics is the belief that we are on the brink of environmental Armageddon, the demise of virtually all higher life forms. If the premise is true then it is plausible to believe that we ought to accept the unsavory consequences of ecocentrism, such as, that circumstances might arise in which an individual snake should be valued more than an individual human. Such a consequence may be distasteful to those still concerned about individual human rights, but not to those concerned about saving life on the planet. And it is the question of whether life on the planet is endangered that is at issue.

Ecocentrists can escape the charge of begging the question, here, and may go on to argue that the long term results of ecocentric practices and institutions would be less misanthropic than the long term results of continuing along our anthropocentric path of resort building. Ecocentrists might argue that anthropocentrism will lead to the death of all life whereas ecocentrism will lead merely to a reduction in the human population level. For these reasons, I think ecocentrism survives these criticisms. It is still a viable, if occasionally brutal, new ethic.

Ecocentrism can withstand the criticisms leveled against it so far. But there are two further questions to be addressed. *(a) Is the scientific foundation of ecocentrism reliable?* And, *(b) Is its ethical method sound?*

(a) Is the science reliable?

The fundamental principle of organismic ecology was expressed by W. A. Allee and coauthors in a text popular in the 1950s.[49] They wrote:

> The [ecosystem] community maintains a certain balance, establishes a biotic border, and has a certain unity paralleling the dynamic equilibrium and organization of other living systems. Natural selection operates upon the whole interspecies system, resulting in a slow evolution of adaptive integration and balance. Division of labor, integration and homeostasis characterize the organism.[50]

The emphasis here upon slow evolution toward an organized state of integrated equilibrium suggests that ecosystems are large scale organisms striving toward a goal. The view that ecosystems are teleological entities with goals is essential to ecocentrism because it establishes that the ecosystem has a good or welfare we can harm or benefit through our actions. An alpine lake, for example, would be teleologically oriented toward the state in which natural selection had operated upon its whole interspecies system to produce an adapted integrated whole consisting of flora and fauna and microorganisms.

The teleological character of ecosystems is crucial to ecocentrism because things that are not internally directed toward a goal do not possess moral standing of their own. Things like tractors, thermostatic heating systems, and computer programs engage in movement, but their movement is directed by external forces. Therefore, their end states are not determined internally, by the things themselves, but rather by agents exerting external influence on the things. It follows that the property we are called upon to respect when we engage in actions that affect tractors and computer programs are the properties, and specifically the interests, of the agents who control the things. The things themselves are not intrinsically valuable.

Another way to put the point is that whereas humans are internally directed, have a good of their own, and therefore are worthy of moral respect in and of themselves, things that lack internal directedness are either directed toward goals that are externally determined, or exhibit an internal structure that appears to be an end of the thing itself but is in reality a *byproduct* of external forces acting on the thing.[51]

If a thing's endpoint is externally determined, then the thing itself does not have intrinsic value. It has instrumental value as the means of some agent toward some goal of the agent's. If a thing exhibits internal structure and integrity that is not the result of its own teleology but is rather the byproduct of forces acting on it, then the thing itself cannot have moral standing because we cannot tell toward what endpoint it tends. Indeed, it does not tend toward any endpoint at all, but rather is directed toward some endpoint by the forces determining its direction.

The paradigmatic case of an internally directed complex system is a person. Persons, or individual humans, can be construed as teleological systems, the aims or goals of which are determined from within the system itself. We call our aims or goals our interests, and because our interests *matter* to us, because we know how it feels to have others frustrate us for trivial reasons in the pursuit of our interests, we attribute to others the basic rights we want to claim for ourselves, rights such as the right to have others refrain from interfering with us, all things being equal, as we pursue our basic interests.

If the good of people can be determined from examining their particular interests, however, the good of specific *parts* of the individual cannot be similarly determined. The good of a finger or a liver or an eye is not determined by finding out toward what endpoint it is internally directed. The good of a bodily organ is determined by finding out what function it plays within the overall organism. What is good or bad for a finger is determined by finding out what hinders or enhances a finger's fitness to play the role fingers are designed to play.

Fingers exhibit an internal complexity of structure in which many parts--blood vessels, muscles, bones--are harmoniously organized to serve the larger organism. But this internal structure must not be mistaken for an end state toward which the finger aims because it is a byproduct of many forces, internal and external to the finger, operating independently of the finger and controlled by overarching forces at the organismic, or individual, level. There is no objective feature or internally decided aim of the finger itself by reference to which we could decide what the good of the finger is. Therefore, whereas individuals are internally directed and have a good of their own, parts of individuals are externally directed and are "good" only in virtue of their serving as fit instruments for some purpose.

Just as parts of individuals have no good of their own, so groups of individuals have no good of their own. My family is a group of individuals with an internal structure and complexity. We might be tempted to say that the group appears to be internally directed toward, perhaps, a mature equilibrium state of harmony and convergence of interests. But my children's interests and my wife's interests do not always converge, nor does there seem to be any outside force which acts externally upon us to forge our various interests into one. Any convergence of interests seems to result from individuals consciously working to sacrifice individual interests that threaten to destroy the happiness of other family members. Any harmony of family life seems to come from each individual nurturing those interests that serve to promote not only their narrow self-interests but the interests of other members as well. Any apparent endstate toward which our family seems to be internally directed turns out, on examination, to be the byproduct of choices and actions of individuals severally pursuing their own good where an important part of their own good includes the happiness of others.

In the early part of the twentieth century, Arthur G. Tansley criticized organismic ecology for construing biotic communities as individuals. Tansley argued that ecosystems are no more organisms than collections of organs are individuals. A human liver or kidney or heart is not capable of existing without the other organs around it. Take a finger off of a human being and it will not be able to grow, because its good is dependent on its being connected to the rest of the human physiological system. The same is not true of individual plants and animals in ecosystems. Take a typical cactus seedling out of its native Arizona ecosystem and try to grow it in inland South Carolina, or Iowa, and it will do just fine.

Gary Varner explains the problem as follows:

> [W]hile ecosystems are sufficiently analogous to organisms to be called "quasi-organisms," the disanalogies show decisively that ecosystems are not organisms in any literal

> sense. In particular, Tansley stressed that, unlike the organs of a body, individual organisms from a given ecosystem are capable of existing independently of each other. From this "general independence" of an ecosystem's constituent organisms follow two other disanalogies between it and an organism: ecosystems lack "the physical unity and definiteness of outline" characteristic of an organism, and an ecosystem's organisms can "transfer themselves to another community and become true members of it," an ability with no significant analog in an organism's organs.[52]

Tansley showed the impropriety of thinking of ecosystems as organisms, an impropriety illustrated by the fact that an ecosystem's constituent parts are not always dependent on the ecosystem in order to survive.

There is a second problem, the difficulty of identifying the boundaries of an ecosystem. In Callicott's example we can ask, What are the boundaries of the hillside ecosystem? We might be tempted to think that this marmot is as important an animal as any in the whole, but sometimes it lives on the hillside and sometimes it does not. Is it in or out? What about this particle of soil? It was not always here, and it may be on the verge of washing down into the stream and out of the system. Call this the problem of spatial identity.

There is also a problem of temporal identity. The species of fish and benthic organisms Callicott wants to protect in the hillside stream were not always here. Go back a few decades and you find white spruce and limber pine growing where maple and oaks now tower. Go back a few hundred years, and the hillside is covered with big bluestem grass, not a tree in sight. Go back a few thousand years, and the climate is so cold that you find nothing but glacial ice scraping over the surface. Go back a hundred thousand years and find tortoises and sharks and trilobites swimming over the hillside in a shallow warm sea.[53]

Through time, a geographical location exhibits many mature equilibrium states, none of which can simultaneously exist with any of the others. But which one is the "natural" one we must respect? If we cannot answer this question, then we cannot identify the standard by which to judge which actions in this geographical location will be morally acceptable.

The general problem indicated by the spatial and temporal fluidity of ecosystems is that unless we can say what a thing's interests are or, at least, toward what end or ends it is internally directed, we cannot say what actions on the parts of moral agents are morally praiseworthy or blameworthy with respect to that thing. If we can say what a thing's interests or aims are, then we can say what actions will promote or thwart that thing's internal directedness. In the case of persons, we can say both what things are good and bad for a person, and what a person desires to do, and therefore we can say what things are right or wrong for agents to do with respect to that person.

But does nature have desires, things it wants to do? Nature knows a vast number of processes which seem to be internally directed in one way or another, but because nature is so wonderfully complex and because things are so intimately interconnected, the appearance of internal directedness may be only an appearance.

Consider the shape and location of a cloud. We can say that *that* cloud has boundaries, complexity, stability and internal direction, but we also know that its stability will be very short lived, its complexity will be very fluid, and the direction predicted for it at this moment will probably be quite different from the direction predicted for it thirty seconds from now. To attribute internal direction to a cloud, then, is to ignore the fact that the identity of the cloud is almost completely determined by outside forces. Were we called upon by moral law to respect the cloud itself, we would be incapable of performing our duty because respecting the cloud would also entail respecting an infinite list of things external to the cloud: the surrounding high and low pressure zones, wind currents in other states, and the geography of landforms beneath it. But if respecting the cloud itself entails respecting all of these other things, in what sense can we respect the cloud itself?

Language confuses us here. What we mean to say is that we must respect the larger system of which the cloud is a part. Ecosystems are like clouds, their boundaries so plastic and their identities so determined by external forces, that it becomes impossible to say how an ecosystem's own identity is distinct from all of the external forces operating on it. Consequently, when we say we must respect ecosystems, we mean we must respect the larger system of which the ecosystem is a part.

Return to agriculture. It is clear that a farmer who clearcuts a slope changes its ecosystem, but has the farmer interfered with the natural progress of an internally directed system? That is, are ecosystems entered by farmers more like human individuals entered by surgeons or more like clouds entered by airplanes? This is an empirical question to be answered by specialists familiar with diverse ecosystems. But the answer probably lies somewhere between the two extremes, as most ecosystems are less coordinated and teleologically structured than human individuals but longer lasting and more stable than clouds. However we decide this question, it seems obvious that farmers tilling prairies will disrupt the natural habits and tendencies of individuals in the ecosystem, and this fact constitutes a reason to be concerned about farming.

But on the other hand, farming will not completely destroy all life in the location. As Hettinger points out in another context, a forest disturbed by a farmer may be "gone, true enough, but micro-ecosystems continue to exist, fireweed and other sun-loving plants will shoot up in the spring forming new biological communities, and in any case, over the long term the forest is likely to return."[54] Biotic communities clearly are not as evanescent as clouds, but neither are they as tightly organized as human individuals. They are complex constantly changing systems. The problem of identity is that the boundaries and "natural" equilibrium states of ecosystems are so difficult to identify.

An organic ecologist might reply that the answer to the identity problem is solved just by specifying a time frame. Just as the fact that a woman's identity changes between her preteen years and her late middle age years without calling into question the fact that she has an identity *today*, so the fact that an ecosystem's identity changes over centuries need not call into question the fact that *it* has an identity today. Carefully define a time frame and we can identity the equilibrium state toward which a given biological whole is tending at that time. Within the equivalent of its own "times of life," each ecosystem exhibits at least a rough stability.

The problem with this response is ecosystems have several "identities" simultaneously, depending on the criteria used to define them. Consider the different ways in which species populations, which are but a single component of an ecosystem, can be identified. There are at least three ways of understanding community, depending upon whether we focus on where things live, what they eat, or how they are biologically related:

> Spatial communities include all of the species within a specific habitat or habitat stratum. Trophic communities include all of the species at one trophic level, all of the species located in a pair of trophic levels, or the "guild" of all species using the same resource. Taxonic communities include all species of some higher taxon.[55]

Communities are only one part of ecosystems, which also contain abiotic elements--such as chemicals and rocks--and natural processes--such as the hydrologic cycle and evolutionary adaptation of species. The problem of definition found with "communities," applies equally to "abiotic elements" and "natural processes." Everything depends on how we draw our boundaries, and we typically draw our boundaries in order to serve some scientific need.

Brian Steverson suggests that the problem is "apparent arbitrariness" in ecosystem identities, an arbitrariness that calls into question the very idea of there being an ecosystem identity:

> If communities are pure constructs, and nothing more, then one can seriously doubt whether they represent any state of affairs inherent in and essential to the natural world. The notion of "community" might be nothing more than a theoretical, heuristic device. If so, then higher levels of ecological organization, such as ecosystems and biomes . . . are also infected with an inherent meaninglessness.[56]

Steverson jumps too quickly from "apparent arbitrariness" to "meaninglessness," but he identifies the problem. Which perspective gives us the "real" communities of an ecosystem? And which perspective gives us the "real" or "natural" ecosystem? Without an answer to this question, we cannot specify which natural state of an ecosystem is the one with which moral agents such as farmers are bound not to interfere.

The identity problem goes deeper. Organic ecology is based on a teleological philosophy of science much like Aristotle's. For Aristotle, there were "natural tendencies" toward which biological individuals and systems were aimed, and there were interfering forces that sometimes prevented these individuals and systems from reaching their desired state. As Sober puts it,

> Heavy objects in the sublunar sphere have location at the center of the earth as their natural state; each tends to go there, but is prevented from doing so. . . . [As Aristotle writes] " . . . for any living thing that has reached its normal development and which is unmutilated, and whose mode of generation is not spontaneous, the most natural act is the production of another like itself, an animal producing an animal, a plant a plant . . ."[57] . . . According to Aristotle, mules (sterile hybrids) count as deviations from the natural state. In fact, females are monsters as well, since the natural tendency of sexual reproduction is for the offspring to perfectly resemble the father . . .[58]

Unlike holistic organismic biology, another form of biology, individualistic Darwinian biology, rejects the natural state model in favor of a model in which variation and mutation is considered the norm. As Sober explains,

> It isn't just that Aristotle was wrong in his detailed claims about mules and women; the whole structure of the natural state model has been discarded. Population biology is not conceptualized in terms of positing some characteristic that

> all members of a species would have in common, were interfering forces absent. Variation is not thought of as a deflection from the natural state of uniformity. Rather, variation is taken to be a fundamental property in its own right.[59]

For population biologists, a "norm of reaction" describes the various forms an individual plant or animal may take, given its particular genotype:

> The norm of reaction of a genotype within a range of environments will describe what phenotype the genotype will produce in a given environment. Thus the norm of reaction for a corn plant genotype might describe how its height is influenced by the amount of moisture in the soil. The norm of reaction is entirely silent on which phenotype is the "natural" one.

As with the relationship between plant genotypes and environment, so with the relationship between population levels and ecosystems. The norm of reaction for a deer population in an ecosystem might describe how the number of deer is influenced by the number of predators in the area, but it will be entirely silent on which number of deer is the "natural" one.

Undermined by scientific observation and theory, organic holistic ecology has been largely replaced by a new paradigm characterizing natural systems in terms of change and disturbance and natural selection at the individual level rather than stability and integrity at the systems level. According to the individualistic model, the integrity and stability of ecosystems is to be explained not in teleological terms of a goal toward which the ecosystem is internally directed but rather in terms of the adaptations of individuals and populations within the system. As Robert E. Ricklefs puts it,

> The ability of the community to resist change [is] the sum of the individual properties of component populations. . . . Relationships between predators and prey, and between competitors, can affect the inherent stability of the community, but trophic structure does not evolve to enhance community stability.[60]

Contemporary population ecologists think of ecosystems not in terms of intentional systems aimed at some future ideal state but rather in terms of individual organisms interacting in ways that can be described using statistical science.[61] The idea that an ecosystem has a "natural" balance which can be harmed or benefited by outside "interfering forces" is, in Sober's words, "entirely alien to post-Darwinian biology."[62] The natural balance of a system is regarded as a byproduct of the many actions of individuals and forces of natural selection, not as the goal toward which a teleological system is aiming.

While some ecologists continue to work with the organismic model, most have largely abandoned it for its tendency to over-emphasize the internal stability and self-integration of ecosystems. This fact does not bode well for any environmental ethic constructed upon it.[63] If individuals in nature do not exist for the sake of the whole—if the whole, whatever it is, exists primarily as a by-product of the properties and actions of individuals in it—then the ideal of "living in harmony with nature" becomes an attractive metaphor. But a metaphor from which it is virtually impossible to derive any specific action-guides.

Let us turn our attention from the scientific foundations of ecocentrism to its philosophical structure.

(2) Is the ethical method sound?

According to Norman Daniels, an ethic is a theory we reach via the method of wide reflective equilibrium, that is,

> a coherent ordered triple of sets of beliefs held by a particular person, namely a set of considered moral judgments (a); a set of moral principles, (b); and a set of relevant background theories, (c). . . . The agent may work back and forth, revising his initial judgments, moral principles, and background theories, to arrive at an equilibrium point which consists of the triple, (a), (b), and (c).[64]

To reason in this fashion is to begin to form moral principles, such as "Always tell the truth," on the basis of compared intuitions. When we hold to such principles we must sometimes revise our intuitions about practical matters because our principles, which cover a wide range of similar cases, bring all of our considered judgments into equilibrium and may tell us that one or another of our original intuitions were wrong. Ethics is about reasoning in this way so that we can come to act on principles that preserve the greatest number of the considered judgments we deem most central to our overall web of values.

So far we have considered only moral knowledge. We can go on to bring knowledge other than moral principles to bear on our practical decisions, knowledge such as that available from ecology, molecular biology, sociology, and political theory. When we bring this background scientific knowledge into our deliberations we are engaged in the full spectrum of wide reflective equilibrium. An ethic, to repeat, is a moral theory in which our (a) considered intuitions have been brought into equilibrium with our (b) moral principles and (c) scientific knowledge.

Considered moral judgments, moral principles and relevant background theories must all be reasonably independent sources of information. If we mistake one for the other, our resulting ethic will not be in equilibrium. If we mistake one of our aesthetic values, for example, for the deliverance of a science, then we will have reasoned in a circular fashion and we will not have arrived at a well reasoned position.

When I first learned that researchers were injecting bGH into dairy cattle, my initial reaction was that it would be morally problematic to inject cows with synthetic chemicals foreign to them, and that it might result in elevated levels of bGH in the milk children drink. However, I learned that scientists, working independently, and using methods that could be replicated and verified by others, had proven just the opposite. bGH occurs naturally in all cows, and the milk of injected cows does not contain any greater amount of bGH than milk from cows not injected with bGH. This scientific information was was independent of my initial moral intuition, and helped me to see that my initial reaction was dead wrong.

Had the scientific community, contrary to fact, declared bGH milk unsafe for human consumption on the basis of defective experiments, then my original moral intuition would have been in equilibrium with my background scientific knowledge. However, if the scientific community, on the basis of valid experiments, reversed itself and declared bGH milk safe, then I could not continue to hold to my original intution. In this fanciful thought experiment, once I learned of the illegitimacy of the science on which the unsafe pronouncement was based, I would be guilty of mistaking my original intuition for a deliverance of science were I to continue to maintain my belief about the dangers of bGH milk.

In ethics we go back and forth between (a), (b), and (c), revising our first intuitions in light of other intuitions, general moral principles and scientific information; then revising our general principles in light of carefully considered intuitions and new scientific information; and revising our scientific theory, every so often, when the number of anomolous observations become so weighty and troublesome that they cause scientists to jettison existing theory in favor of a new one. But there is a danger in moral reasoning to which we must be alert: to confuse circular reasoning, an unsound method, with the sound method of comparative, back and forth, reasoning toward reflective equilibrium.

Many of us who came of age after the first Earth Day share the intuition that nature is composed of intrinsically valuable biological wholes. But ecocentrism seems to conflate this intuition with science. The background scientific theory of Callicott's ecocentrism is evolutionary theory, sociobiology, and a naturalistic moral psychology. To quote from a passage discussed above, the logic of ecocentrism

> is that natural selection has endowed human beings with an affective moral response to perceived bonds of kinship and community membership and identity; that today the natural environment, the land, is represented as a community, the biotic community; and that, therefore, an environmental or land ethic is both possible . . . and necessary . . .

There are two appeals to science here. The second appeal contends that scientists see the land as an organic community. We have already discussed reasons for doubting this claim, reasons strong enough to justify us in wondering whether the appeal to the communal quality of the land is more a *presupposition* of a certain kind of ecological science than a result of experimental inquiry. Callicott's asserting that land is a biotic community seems to beg the question about whether an ecosystem's mature equilibrium state is best considered a goal of a single organism or a byproduct of many individuals each severally pursuing their own goals.

The first appeal to science contends that science shows that morality evolves from natural selection and that our values are based therefore on natural affections and emotions. Callicott here gives a Darwinian turn to David Hume's moral theory, arguing that moral sentiments are best explained in terms of selective adaptations. But this argument, too, is not the conclusion of any scientific investigation but rather a philosophical judgment based on a certain view of humans and dependent upon the presuppositions of sociobiology. Sociobiology is a social science with strong competitors, competitors that insist that an etiological explanation of the history of morality is not the same as a philosophical justification of morality.

Ecocentrists start from a set of aesthetic values about the integrity, stability and beauty of the environment and a sense of moral outrage at the degradation inflicted on the land by modern agriculture and human overpopulation. They consider and refine these intuitions into moral principles, namely, that nature is intrinsically valuable and that farming is intrinsically destructive. Background theories such as holistic ecological science cohere with these intuitions and principles and lead us to believe that our original intuitions are in wide reflective equilibrium with background scientific knowledge.

On examination, however, the ecocentrist's chosen moral principles do not explain all of our intuitions, particularly the intuition that every human being has a right to be fed if there is food enough to go around. To the extent that we have any obligations at all to other humans, it would seem to be one of our basic duties, commensurate with others' basic moral rights, that we endeavor to feed the hungry. To abandon the arts of cultivation is not to endeavor to feed the world's hungry, of which there are now over two billion. The justification of the practice of agriculture, then, is secured by whatever arguments justify the existence of our most basic duties to others. To argue that agriculture itself is morally unjustifiable is to assume an onerous burden of proof.

If the science to which ecocentrism appeals to justify its original intuitions converges too neatly with those initial intuitions, then the ecocentric arguments offered on behalf of environmentalist intuitions and moral principles may be circular. Ecocentrists may be mixing up the results of environmental science with their own aesthetic preferences, mistaking intuitions about the intrinsic value of nature for the deliverances of ecological science.

We have spent much time investigating the scientific and philosophical foundations of ecocentrism because it is the strongest theory available for justifying the attribution of intrinsic value to nature. We have seen that there are reasons to doubt the scientific foundations of ecocentrism and to be skeptical about its philosophical structure. Therefore, ecocentrism does not provide us with the theory we need to justify belief in **(14.2)**. In the absence of another theory justifying belief in the thesis that nature is an internally-directed individual with goals of its own, the idea that we could vex nature no longer seems compelling.

Conclusion

Many of the intrinsic objections to ag biotech protest that ag biotech is unnatural **(UE)**. As Hume pointed out in the eighteenth century, and as we noted in our discussion in ch. 2 of the is/ought fallacy, there are good reasons to be cautious when trying to argue moral matters in terms of what is natural. It is notoriously difficult to derive valid normative statements from empirical claims. The fact that the world is set up in a certain way (a matter to be expressed in empirical terms) is not necessarily a good reason to believe that it ought to be set up that way (a matter to be expressed in normative terms). For it is empirically true that there are racist societies and yet this fact does not supply a good reason to believe that there ought to be racist societies. It is a fact that some children torture cats. Yet this fact does not supply a good reason to believe that children should be cruel to animals. What is the case is logically distinct from what ought to be the case.

We should not ignore the world in arriving at our moral views, but we ought to be careful in claiming that Mother Nature gives final guidance.

What appears to us to be the very essence of nature may be little more than our own prejudices read *onto* nature.

The point is brought home by considering the mistake Aristotle made in his Natural Law theory. When Aristotle looked at nature he did not see great variability, but rather a small number of essences, from which there were deviations. As Sober observes, Aristotle distinguished

> between the natural state of a kind of object and those states which are not natural. These latter are produced by subjecting the object to an interfering force. In the sublunar sphere, for a heavy object to be in its natural state is for it to be located where the center of the Earth is now (*On the Heavens*, ii, clr, 296b and 310b, 2-5). But, of course, many heavy objects fail to be there. The cause for this divergence from what is natural is that these objects are acted on by interfering forces which prevent them from achieving their natural state by frustrating their natural tendency. Variability within nature is thus to be accounted for as a deviation from what is natural . . . [65]

An essentialist theory, or Natural State model, neither fits contemporary scientific observations nor satisfies current moral sensibilities. Others have shown its scientific weaknesses.[66] For our purposes, we must be aware of its powerful tendency to reinforce morally jaundiced views.

When we think that one instantiation of an organism is the essential or natural state of many different organisms, then it is easy to think that organisms unlike the favored instantiation are deviant. Aristotle considered entire species unnatural because they were unlike what he took to be the "normal" state. As he wrote in the *Generation of Animals*:

> Seals are deformed as a group because they resemble lower classes of animals, owing to their lack of ears. Snails, since they move like animals with their feet cut off, and lobsters, because they use their claws for locomotion, are likewise to be counted as monsters (*Generation of Animals*, 19, 714b, 18-19; *Parts of Animals*, iv, 8 684a35).[67]

Aristotle also considered some groups of people to be inferior, believing that some were by nature fit to be slaves, and that women were not owed the kind of regard owed to Greek men. Why? The Natural State model appears in the justification: Nature simply had not seen fit to equip non-Greeks with the virtues of the Greeks, or women with the rational capacities of men. Therefore, it would be unnatural for non-Greeks to have the rights of civilized peoples, or for women to make decisions in the polis.

Surely we ought to reject such moral views, however. And we similarly ought to be wary of theories that take nature to be the primary source of correct moral opinions. To view nature in this fashion is to run the risk not only of failing to recognize one's own moral biases, but to baptize them with an honorific title "natural." In a culture in which natural is good and unnatural is bad, calling an action unnatural may have the effect of reinforcing prejudices we ought to have abandoned.

We have examined fourteen intrinsic objections to ag biotech. None of these objections seems sound, save perhaps for **(10)**. But **(10)** proves too much for anyone willing to accept technological intervention in sexual reproduction to help childless couples conceive. Henceforth, critics of ag biotech must either bring forth other intrinsic objections, or focus on ag biotech's potentially adverse consequences. To which we now turn.

Notes

1. Stephen Stich, "Forbidden Knowledge," in R. P. Bareikis, ed. *Science and the Public Interest: Recombinant DNA Research* (Bloomington, IN: Indiana University Foundation, 1978).
2. Duvick's research at Pioneer focused on the developmental cytology and biochemistry of the maize endosperm.
3. Duvick's lecture was later published as "Is Biotechnology Compatible with Sustainable Agriculture? Yes," in *The Ag Bioethics Forum* 4 (June 1992). Also see his articles, "Biotechnology is Compatible with Sustainable Agriculture," *Journal of Agricultural and Environmental Ethics* 8 (1995): 112-125; and "Our Vision for the Agricultural Sciences Needs to Include Biotechnology," *Journal of Agricultural and Environemtal Ethics* 4 (1991): 200-206.
4 Mae-Wan Ho, *Genetic Engineering Dreams or Nightmares? The brave new world of bad science and big business,* Third World Network and Research Foundation for Science Technology and Ecology, New Delhi, India, 1997. Christian Aid, "Selling Suicide: Farming, False Promises and Genetic Engineering in Developing Countries," on-line at: www.christian-aid.org.uk/reports/suicide/index.html
5 Cf. Carolyn Raffensperger and Joel Tickner, eds., *Protecting Public Health and the Environment: Implementing the Precautionary Principle* (Island Press, 1999).
6 K. N. Stauber, C. Hassebrook, E. A. R. Bird, G. L. Bultena, E. O. Hoiberg, H. MacCormack, and D. Menanteau-Horta, "The promise of sustainable agriculture," in E. A. R. Bird and J. C. Gardner, ed., *Planting the future: Developing an Agriculture that Sustains Land and Community* (Ames, IA: Iowa State University Press, 1995), pp. 3-15. Cited in Robert L. Zimdahl, "Moral Certainty in Agriculture," unpublished manuscript, p. 4.
7. Luther Tweeten, "Will the Benefits and Costs of Bovine Growth Hormone Be Distributed Fairly?" paper presented at conference on "Ethics and Agricultural Biotechnology," at Iowa State University, Ames, May 22, 1991, p. 16.
8. According to Gary Jolliff, about fifty percent of this increase can be attributed to the work of plant breeders. Jolliff, "Strategic Planning for New-Crop Development," *Journal of Production Agriculture* 2 (January-March 1989): 8.
9. John E. Lee, Jr., and Gary C. Taylor, "Agricultural Research: Who Pays and Who Benefits?" in John J. Crowley, ed., *Research for Tomorrow: 1986 Yearbook of Agriculture* (no publication information given, no date): 16-17. William McNeill observes that "taller stature offers definite proof of the superior nutrition Americans began to enjoy as early as 1700" ("Gains and Losses," p. 17).
10 Among others who have made this distinction are Michael Reiss and Roger Straughan, *Improving Nature: The Science and Ethics of Genetic Engineering* (Cambridge: Cambridge University Press, 1996), pp. 49-51.
11 Baruch Brody, private communication.
12. Because of the close ties between humans and nature in this paradigm, some environmental philosophers come to its defense. J. Baird Callicott, for example, suggests that we ought to recover large portions of it, citing with approval Aldo Leopold's remark that humans should

"get their meat from God," meaning that we should only consume meat that comes from wild beasts we have actually hunted and gutted, and only eat nuts and berries we have discovered and gathered for ourselves.J. Baird Callicott, *In Defense of the Land Ethic: Essays in Environmental Philosophy* (Albany: S.U.N.Y. Press, 1989), p. 36, quoting Leopold's *Sand County Almanac* (New York: Oxford U. Press, 1949), p. viii and p. 166.
13. Cf. G. H. Fairtlough, "Genetic Engineering--Problems and Opportunities," in S. Jacobsson, A. Jamison, and H. Rothman, eds., *The Biotechnological Challenge* (Cambridge: Cambridge University Press, 1986): 17.
14. William McNeill, "Gains and Losses: An Historical Perspective on Farming," The 1989 Iowa Humanities Lecture, National Endowment for the Humanities and Iowa Humanities Board, Oakdale Campus, Iowa City, Iowa, p. 5.
15. Richard L. Willham, "The Legacy of the Stockman," Department of Animal Science, Iowa State University, Ames, Iowa, 1985, p. 61.
16. Stephen Jay Gould, "On the Origin of Specious Critics," *Discover* (January 1985): 34-42.
17. Steven Burke, private communication, 6 April 1993. Quoted with permission.
18. Elliott Sober, "Evolution, Population Thinking, and Essentialism," *Philosophy of Science* 47 (1980): 351.
19. Bernard Rollin, *The Frankenstein Syndrome: Ethical and Social Issues in the Genetic Engineering of Animals* (New York: Cambridge University Press, 1995), p. 159.
20. Rollin, *Frankenstein*, 192.
21. Bill Muir, email correspondence. See reports of his research in *Poultry Science* 75 (1996): 294-302, and 447-458.
22. Rollin, *Frankenstein*, 192.
23 Dr. J. Gressel made this observation in orally after this paper was presented at the annual meeting of the Weed Science Society of America.
24 Cf. A. Goude, *The Human Impact: Man's role in Environmental Change* (Cambridge, MA: MIT Press, 1981), p. 67.
25. Cf. Dorothy Nelkin and M. S. Lindee, *The DNA Mystique: The Gene as Cultural Icon* (New York: Freeman, 1995).
26. OTA, Patenting Life, p. 31.
27 "Where is the Holy Spirit Anyway? Response to a Skeptical Environmentalist" *Ecumenical Review* 42 (April 1990): 165.
28. One of the legacies of Leopold's *A Sand County Almanac* was its call for a new "land ethic." *A Sand County Almanac: With Essays on Conservation from Round River* (New York: Ballantine Books, 1966, originally published in 1949), esp. pp. 237-263. As Callicott points out in *In Defense of the Land Ethic: Essays in Environmental Philosophy* (Albany, NY: S.U.N.Y. Press, 1989), p. 278, n. 4, Leopold first published his ideas as "The Conservation Ethic," in *the Journal of Forestry* 31 (1933): 634-643.
29. Leopold, *Almanac*, 238.
30. Richard Sylvan (Routley), "Is There a Need for a New, an Environmental, Ethic?" *Proceedings of the XV World Congress of Philosophy*, No. 1. Varna, Bulgaria, 1973, pp. 205-210; reprinted in Michael E. Zimmerman, ed., *Environmental Philosophy: From Animal Rights to Radical Ecology* (Englewood Cliffs, NJ: Prentice Hall, 1993), pp. 12-21.
31. For an ecofeminist ethic, see Karen J. Warren, "The Power and Promise of Ecological Feminism," *Environmental Ethics* 12 (Summer 1990): 125-146.
32. For one of the first statements of a biocentric view, see Kenneth E. Goodpaster, "On Being Morally Considerable," *The Journal of Philosophy* 75 (June 19789): 308-325, reprinted in

Michael E. Zimmerman et al, pp. 49-65. The definitive exposition of a biocentric ethic is Paul Taylor, *Respect for Nature* (Princeton, NJ: Princeton University Press, 1986).
33. Rodman, "Four Forms," p. 89.
34. Callicott seems to prefer the term "land ethic," but writes that he counts himself an ecocentrist. *Defense*, p. 3.
35. Callicott, *Defense*, 2-3.
36. Leopold, "The Land Ethic," in *A Sand County Almanac* (New York: Oxford University Press, 1949), pp. 224-225.
37. Holmes Rolston, III, *Philosophy Gone Wild: Essays in Environmental Ethics* (Buffalo: Prometheus Books, 1986): 18.
38. Callicott sees his views as the logical development of Leopold's ideas. I believe Paul Taylor's biocentric ethic is equally as mature a position, and equally as deserving of comment, as Callicott's, but Taylor's ethic is individualistic and preserves the "old and prevailing" idea of individual rights. Therefore, it does not meet Sylvan's criterion of being radically different from prevailing ethical theories.
39. Callicott, "Conceptual Foundations," p. 83.
40. Ned Hettinger and Bill Throop, "Can Ecocentric Ethics Withstand Chaos in Ecology?" Draft, April 22, 1994. Quoted with permission.
41. Leopold, Sand County Almanac, p. 245; quoted by Sylvan on p. 13.
42. Callicott, Land Ethic, p. 22.
43 At a conference of the Society for Food, Agriculture, and Human Values, in Tuscon, Arizona, in 1994, I heard Angelo Joaquin discuss the Tohono O'odham, Desert People. The Tohono O'odham have a remembered history of the place; they know its past, its possibilities, its strengths, its limits. Their indigenous knowledge and uncanny ability to flourish on prickly pear fruit, mesquite beans and teperee beans reveal that they are, in Joaquin's apt phrase, a people "meant to live here, a people who know how to eat from the desert's bounty."
44. For example, Tom Regan, William Aiken, and Elliot Sober. William Aiken, "Ethical Issues in Agriculture," in Tom Regan, ed., *Earthbound: New Introductory Essays in Environmental Ethics* (New York: Random House, 1984), p. 269; Tom Regan, *The Case for Animal Rights* (Berkeley: University of California Press, 1983); Elliot Sober, "Philosophical Problems for Environmentalism," in Brian Norton, ed., *Preservation of Species* (Princeton: Princeton University Press, 1986), pp. 173-194.
45. Edward Abbey, *Desert Solitaire* (Tucson: The University of Arizona Press, 1968), p. 33.
46. "Some indication of the genuinely biocentric value orientation of ethical environmentalism is indicated in what otherwise might appear to be gratuitous misanthropy." *Defense*, p. 27.
47. Callicott, *Defense*, p. 27.
48 Regan, *Case*, P. 362. Cf. William Aiken, "Ethical Issues in Agriculture," in Regan, *Earthbound,* p. 269; and Marti Kheel, "The Liberation of Nature: A Circular Affair," *Environmental Ethics* 7 (Summer 1985): 135.
49 According to Harley Cahen, "Against the Moral Considerability of Ecosystems," *Environmental Ethics* 10 (Fall 1988).
50. W. A. Allee et al., *Principles of Animal Ecology* (Philadelphia: W. B. Saunders, 1949), p. 728; cited in Cahen, "Against," p. 210, n. 69.
51. I am indebted to Cahen for this crucial distinction between integrated states of affairs that are byproducts and those that are aims.
52. Gary Varner, "In Nature's Interest? Interests, Animal Rights, and Environmental Ethics," unpublished manuscript, dated 23 February 1993, p. 1.11. Quoted with permission.
53. Hettinger and Throop, p. 6. The authors cite "Botkin, Discordant Harmonies, Ch. 3."

54. Hettinger and Throop, p. 7.
55. Brian K. Steverson, "Ecocentrism and Ecological Modeling," *Environmental Ethics* 16 (Spring 1994): 77.
56. Steverson, p. 77.
57. Sober, "Evolution," p. 183. The quotation is from Aristotle's De Anima, 415a26.
58. Sober, p. 183.
59. Sober, p. 183.
60. Robert E. Ricklefs, *The Economy of Nature* (Portland, OR: Chiron Press, 1976), p. 355; cited in Cahen, p. 210.
61. As Cahen reminds us, Donald Worster made this criticism of ecocentrism in 1985 in *Nature's Economy: A History of Ecological Ideas* (Cambridge: Cambridge University Press, 1985), pp. 332-33. Cahen's article has deeply influenced my own views on this matter.
62. Sober, "Evolution," 183.
63. Cahen identifies two who accept group selection at the community level: Eugene Odum, *Fundamentals of Ecology, 3rd ed.* (Philadelphia: W. B. Saunders, 1971), pp. 251-75; and M. J. Dunbar, "The Evolution of Stability in Marine Environments: Natural Selection at the Level of the Ecosystem," *American Naturalist* 94 (1960): 129-36. See Cahen, p. 215.
64. Norman Daniels, "Reflective Equilibrium and Archimedian Points," *Canadian Journal of Philosophy* 10 (March 1980); 85-86. Daniels' account of how we construct an ethic is based on work of John Rawls.
65 Sober, "Evolution," 360.
66 For essentialists, "somewhere within the possible variations that a species is capable of, there is a privileged state--a state which has a special causal and explanatory role. The laws governing a species will specify this state...[but] the diversity of individual organisms is a veil which must be penetrated in the search for invariance" (Sober, p. 364-5). Modern biologists reject the natural state model because individuals within species have neither identical phenotypes nor genotypes. As Sober puts it, "diversity itself constitute[s] an invariance, obeying its own laws" (p. 365).
67 Sober, "Evolution," 362.

Chapter 6

Problems for the Case Against Ag Biotech, Part II: Extrinsic Objections

(a) If there is a substantial risk that a technology will do more harm than good to humans, ecosystems, and animals, then it should not be developed.
(b) There is a substantial risk that ag biotech will do more harm than good to humans, ecosystems, and animals.
(c) Therefore, ag biotech should not be developed.

Thus the extrinsic argument against ag biotech. In considering whether it is a good argument, I believe we should simply assume that *(a)* is true. *Substantially* risky technologies, perhaps by definition, should not be developed. Seeing no reason to contest *(a),* therefore, I will focus on *(b)*.

Eight extrinsic arguments

There are at least eight arguments in support of *(b)*.

(1) There is a substantial risk that ag biotech will do more harm than good to humans, by introducing genetically engineered foods carrying unacceptable risks to *human health*.
(2) There is a substantial risk that ag biotech will do more harm than good to humans by perpetuating social inequities in developed economies where it will lead to advantages for larger agribusiness farmers that will be unjustly denied to smaller *family farmers*.
(3) There is a substantial risk that ag biotech will do more harm than good to *subsistence farmers*, by perpetuating social and economic inequities between developed economies with their well capitalized farmers, and developing economies, with their under capitalized farmers.
(4) There is a substantial risk that ag biotech will do more harm than good to *scientists and consumers,* because scientists must increasingly pursue reductionistic, short term, applied science to

benefit private corporations, rather than holistic, long term, theoretical science to benefit public taxpayers.

(5) There is a substantial risk that ag biotech will do more harm than good to *future generations* **by foreclosing possibilities for them to feed themselves.**

(6) There is a substantial risk that ag biotech will do more harm than good to *ecosystems*, **by leading to environmental catastrophe through release into the wild of virulent genetically modified** *organisms, plants, and fish*.

(7) There is a substantial risk that ag biotech will do more harm than good by narrowing *plant germplasm diversity*, **and more harm than good to the atmosphere by reducing the quality of** *air, soils, and ground and surface waters*.

(8) There is a substantial risk that ag biotech will do more harm than good to research *animals*, **livestock and wildlife, by causing them to suffer or die, or to prevent them from continuing as a species.**

These objections require that we assess the potential consequences of ag biotech. Unfortunately, we lack much of the data necessary to make these assessments and, as Kristin Shrader-Frechette has argued, we have a tendency to misjudge risks even when we have full information.[1] Humans generally are more averse to risks from new technologies than old technologies, even when the risks of the novel technologies are demonstrably less than the risks of the current ones.[2] In the face of risk-aversion and factual uncertainty, we must nonetheless do our best to assess extrinsic objections to ag biotech.

Begin with risks to *humans* identified in **(1)** through **(5)**.

(1) There is a substantial risk that ag biotech will do more harm than good to humans, by introducing genetically engineered foods carrying unacceptable risks to *human health*.

Ten years ago, assessing the truth of this claim would have been very difficult because we then had virtually no wide-scale experience eating commercially grown GMOs. But in the year 2000, we have in the United States more than a dozen years of experience with field trials of GMOs, and four years of experience of eating GM foods on a widespread basis.[3] And, as the years go by, it seems that the mantra of the biotech industry has proven itself to be true: There have to date been no verified reports of virulent GM cells, organisms, viruses, or plant or animal foods having harmed any consumer.

There have been two near-misses.

One: Nuts in soybeans. The largest privately owned seed corn company in the world, Pioneer Hi-Bred International, did research in the mid-1980s to transfer a gene from the Brazil nut into soybeans. They succeeded in producing a soybean with higher efficiency and nutritional content for animal feed. However, during tests of the new bean product in which the skin of individuals known to be allergic to Brazil nut were given skin-pricks, the GM soy-nut product caused an allergic reaction. University of Nebraska scientists demonstrated that the Brazil nut soybean contained the potential to produce deadly reactions in people allergic to nuts.[4]

Had Pioneer produced the bean commercially, it would have been the first time a known allergen had been bred into a food that previously did not contain the allergen in question. Potentially, at-risk individuals could have bought and eaten soy products without knowing that they were also consuming nut proteins that could kill them. However, Pioneer shut down the research project after learning of the results of the allergenicity study, and did not bring the product to market.

Two: Tryptophan. During the 1980s, sales of synthetic compounds promising effortless sleep and relief from premenstrual syndrome soared. One such product contained the chemical l-tryptophan. L-tryptophan is an amino acid essential to the human body available only from food sources, such as meat and dairy; the body cannot manufacture it by itself.

We need tryptophan to help us get to sleep, but foods containing it are high in calories and cholesterol. Americans began cutting their intake of such foods in the 1970s, and sleep problems became common. By 1989, scientists had shown that tryptophan produces a sleep enhancing effect and may also be effective in relieving stress and depression. By the end of the decade, health food stores were selling more than $100 million per year of tryptophan in a non-prescription pill as a nutritional supplement. To meet the new demand, Showa Denko, the world's leading manufacturer based in Japan, ramped up its manufacturing process. The new process involved genetic engineering.

Reports of medical problems soon began to appear. In 1989, a connective tissue disorder called eosinophilia-myalgia syndrome (EMS) was linked to Show Denko's product and, in the next two years, at least fifteen hundred people became ill.[5] Some were partially paralyzed. Thirty-eight died.

The company assumed its new GM product was safe because its earlier, non-GM, variety was safe. Was this assumption justified? Ag biotech critic John Fagan thinks not, speculating that

> the genetic manipulations had increased tryptophan production so greatly that the concentrations of tryptophan within the bacteria had reached very high levels. As a result, the tryptophan and its precursors began to react chemically producing unexpected toxic compounds. To date this company has paid over $1 billion in damages, and litigation is still in progress.[6]

Was tryptophan the first genetically engineered disease? Is it only the first of many unintended side-effects so many fear from GM foods?

The answer seems to be no. The poisonous tryptophan was probably not rendered lethal by the processes of genetic engineering. The culprit seems rather to have been more mundane industrial processes. As Fagan's own book, which appeared in 1995, attests, Showa Denko asked the US Food and Drug Administration to test their original batch of GM tryptophan. This batch was produced using an organism called *Bacillus amyloliquefaciens strain III*. Strain III tryptophan passed the FDA standards and was approved for sale. No EMS cases were traced to this GM variety.

But, when Showa Denko later revised production procedures, they made three changes. First, they began to use another strain of GM Bacillus, strain *V*.[7] Second, they cut in half the amount of powdered activated charcoal used in one of the filtration steps. Third, they allowed some of the product to circumvent altogether another filtration step, called reverse-osmosis-membrane.[8]

One of these three changes was probably responsible for the deaths of consumers, all of whom, apparently, were killed by a batch produced with strain *V*. As Raphals notes in a *Science* article, the deadly contaminant probably arose as a result of the company's cutting corners in the manufacturing process. Raphals speculates that "inadequate filtration might have allowed impurities to pass through." But even if we should blame the deaths on the fact that strain *V* was genetically engineered, this fact in itself does not lead to a condemnation of all genetic engineering, since the original, strain *III*, trytophan--also a genetically engineered form--was apparently safe.

In sum, it is far from clear that tryptophan constitutes a case of food genetic engineering harming consumers.

Let us attempt to set the risks of eating GM foods in the context of the risks associated with eating regular, non-GM foods. Consider sweet corn.

Sweet corn can easily acquire molds both before and after harvest. Moldy corn can have high levels of toxins called fumonisins if the mold is the fungus, *Fusarium moniliforme*. These toxins can cause cancer in rats, pulmonary edema in swine, equine leukoencephalomalacia in horses, and are suspected of causing esophageal cancer in humans.[9] These are risks associated with eating moldy sweet corn. But some forms of genetically modified sweet corn accumulate less fumonisins than other varieties when infected by the fungus.

Bt corn is best known in the popular press because a well-known scientific "comment" paper published in the journal *Nature* suggested that Bt corn pollen kills monarch butterfly larvae. Much attention has been focused on potential environmental harms from Bt corn. As the previous paragraph suggests, however, there is a comparatively neglected potential health *benefit* of Bt corn. A study by Gary Monkvold at Iowa State University suggests that Bt sweet corn is actually *safer* for consumers than present non-Bt varieties, because the modified corn seems to be less susceptible to mold. Here, then, is a GM food that *decreases* one health risk carried by its non-GM cousin.

Consider, too, that some foods contain rather high levels of naturally-occurring chemicals that genetic modification might be able to reduce. For example, solanaceous crops, such as potatoes and green peppers, naturally contain rather high levels of chemicals called glycoalkaloids. These glycoalkaloids protect the vegetables from insects in a way similar to the modes of action of two classes of manufactured insecticides, carbamates and organophosphates.

The impact of glycoalkaloids, carbamate insecticides, and organophosphate insecticides on the nervous system of an insect occurs through the chemical's inhibition of a substance called acetylcholinesterase (AC-ase) in the insect's body.[10] Without the production of AC-ase, the insects nerves fire willy-nilly, causing the insect to lose control over its nervous system.

Just as it is a bad thing for an insect to have its AC-ase production inhibited by natural insecticides, so is it also a bad thing for humans, because our nervous systems function almost identically to insects' nervous systems. Although an insect may eat a large proportion of its body weight from a plant containing glycoalkoids in a short period of time, and perhaps be affected by it, humans very rarely eat enough of a naturally occurring AC-ase inhibitor in a normal foodstuff to be harmed (although humans and animals can be harmed by the very high levels in a close relative of potato, the deadly nightshade). Thus, although the US government regulates the allowable residues of manufactured insecticides in foods, it does not regulate the concentrations of natural glycoalkaloids in foods like potatoes and peppers, which have been part of the healthy human diet for centuries.

The ability of potatoes to make their own chemicals to resist insects does permit an interesting thought experiment, however. Suppose we want to develop a new variety of apple, one that will make its own insect-resisting chemicals and thus not need to be sprayed as often with manufactured insecticides. We go on a world-wide search for a close relative of cultivated apple that is able to produce glycoalkaloids and, in our thought experiment, we find it in a rare crabapple. However, we realize that a breeding program to cross the unusual glycoalkaloid-making crabapple with a commercial apple variety will take years.

On the other hand, we know that potato makes the same chemical as our fanciful crabapple. We have eaten potatoes safely for centuries, and there is no need to regulate potatoes for glycoalkaloid content, even though we know that the concentrations of the chemicals can be quite high. For example, a child who eats a large order of French fries from a typical fast-food restaurant may get many times the maximum dose considered to have no effect on rats in lab experiments.[11]

Currently available techniques might allow us to select the gene for glycoalkaloid production from potato cells and transfer it to apple cells, and subsequently develop a new insect-resistant apple variety years sooner than the conventional breeding program would allow. Moreover, the chemical in the apple would be the same chemical that is in our potatoes.

Which new apple variety should be considered "safer?" Should that determination be based on how we bred the new trait into the variety? Or should it be based on the concentration of glycoalkaloids in the plant material to be consumed?

In this hypothetical example, the latter considerations would be relevant under the current regulatory framework the US government uses for agricultural biotechnology (if, as contrary to fact, the government regulated natural glycoalkaloids). Under the current regulatory framework, consideration of the source of the glycoalkaloid gene and how it was incorporated into an existing apple variety would not be considered of consequence, assuming all other characteristics of the apple were unchanged.

This approach to regulating ag biotech reflects a concern with producing safe and nutritious foods independent of the method used to produce them. To date, this approach has served the public well for, as previously noted, there is no verified example of an ag biotech product that has worked its way through the regulatory process only to cause harm to human health or safety. This is strong evidence that the process of modifying crops with biotech methods is no less safe than the process of plant breeding.

Why is the safety record clean to this point? The answer probably has most to do with the protectionary oversight measures that are in place in the

US. Changing the physiology or biochemistry of a food introduces the possibility of increasing levels of toxic agents in the food. In the United States, responsibility for monitoring these changes has fallen to three federal agencies: the Food and Drug Administration, the US Department of Agriculture, and the Environmental Protection Agency. A word about each one is in order.

The US Food and Drug Administration (FDA) has regulatory authority under the Federal Food, Drug, and Cosmetic Act (FFDCA) to insure the safety and labeling of the nation's drug and food supply, excluding meat and poultry, and including veterinary drugs. GM foods are subject to regulation under FDA provisions that prohibit unsafe adulteration of whole foods (e.g., fruits, vegetables, grains), and untested food additives. The FDA decided that GM foods chemically indistinguishable from foods generally regarded as safe (GRAS) need not be singled out for special testing.

The FDA must, however, evaluate GM foods whenever the basic characteristics of the food have been changed. This rule applies whether the change is intended or not, and covers new substances in the food, such as new fatty acids, carbohydrates, and proteins. If any known toxicants, important nutrients, allergens, or antibiotic resistance selectable markers are introduced into, or if any such previously substances in the food are modified, then the FDA must conduct testing of the GM food.

The USDA is required by existing statutes to regulate certain products of agricultural biotechnology. This includes determining the risk associated with the approval of releasing a product of agricultural biotechnology to the field, either for testing or commercial use. The USDA's testing requirements, as stated earlier, are based on the characteristics of the product rather than its origin as a product of biotechnology. Like all Federal agencies, the USDA must allow public participation in the form of comments regarding the rules that are proposed and used to regulate the approval process for GMOs. Furthermore, for petitions to commercialize a GM product (i.e., remove it from regulated status), the public is notified and invited to comment.

The USDA Animal and Plant Health Inspection Service (APHIS) has regulatory authority under Federal Plant Pest Act (FPPA), and the Plant Quarantine Act.[12] APHIS is charged with protecting American agriculture against pests and diseases, and anyone wishing to conduct a field test of a GM crop, import a GM plant into the US, or move a GM plant from one state to another, must obtain a permit from APHIS.

To obtain a permit to field test a GM crop, one must supply, among other things: complete biological information, including descriptions of all new genes, products, and their origins; the purpose of the test; details about the experimental design; and explanation of the precautions taken to prevent escape of pollen, plants, and plant parts.

The FDA and USDA are not the only agencies responsible for assessing the risks of GM crops and foods. In 1986, Congress established the "Coordinated Framework for Regulation of Biotechnology," and it assigned the US Environmental Protection Agency (EPA) regulatory authority under three statutes: Toxic Substances Control Act (TSCA); the Federal Insecticide, Fungicide, and Rodenticide Act (FIFRA); and the Federal Food, Drug, and Cosmetic Act (FFDCA), Section 408.[13] EPA is charged with insuring the safe use of novel microorganisms and pesticides engineered into crops and with regulating all GM plants producing pesticidal substances not produced by their non-GM cousins.

It is worth noting that the regulatory framework in the US establishes an arm's length between the regulating agencies and the companies being regulated. In addition to these agencies, the USDA sponsors a Standing Committee on Biotechnology in the National Academy of Sciences, which comments to the Secretary of Agriculture on important topics such as the way in which long-term monitoring of GM crops should be pursued and potential risks that ought to be regulated, as, for example, for new pharmaceuticals and antibiotics produced from GM plants.

While a reasonably trustworthy regulatory framework seems to be in place in this country, there are good reasons for continued vigilance and even revisitation of some decisions. In 2000, the National Academies of Science (NAS) National Research Council sponsored a committee to review the issue.[14] The committee reported that it was not aware of any evidence suggesting foods on the market today are unsafe to eat as a result of genetic modification. And it said that no strict distinction exists between the health and environmental risks posed by plants genetically engineered through modern molecular techniques and those modified by conventional breeding practices.

In the judgment of the chair of the committee, the agencies have "generally done a good job." On the other hand, "given the current level of public concern," entomologist Perry Adkisson wrote that "the agencies must bolster the mechanisms they use to protect human health and the environment." In particular, the report questioned the EPA's decision in 1994 to exempt viral coat proteins from regulation. Viral coat proteins are protein shields manufactured by viruses to protect the virus from invaders. Scientists put viral coat proteins into plants and the proteins protect the plants from viral infection. There is evidence that these coat protein genes may be taken up by unrelated viruses that infect the plant, inducing a kind of accelerated evolution which changes the coat structure. Recombination between the inserted gene and the unrelated viruses could create an entirely new kind of virus with potentially grave consequences for agriculture.[15]

These concerns led Lynn Goldman, formerly Assistant Administrator of the EPA's Office of Prevention, Pesticides and Toxic Substances, to declare that she was "humbled" by the NAS report, and to offer that some of the exemptions allowed by the EPA should not have been allowed, including the exemption for viral coat proteins. Goldman further expressed dissatisfaction with the ability of the Coordinated Framework to protect against all allergens in GM foods. Her reason was that we do not know the sequence of most allergens and, therefore, cannot know with certainty whether a new allergen has been produced by genetic modification.[16]

There are good reasons for the USDA, EPA, and FDA to continue to improve procedures and review the adequacy of current GM safety regulations. While the EPA proposed regulations in the late 1980s, the agency still has not formally issued the regulations. So, the NAS report might provide the political platform needed by the EPA to correct the viral coat protein exemption and formally issue the regulations.

Federal agencies have also been criticized for not making their procedures more open to the public, and there is more to be done to maximize transparency.[17] Perhaps as a step in this direction, the Secretary of Agriculture has established an Advisory Committee on Agricultural Biotechnology. The Committee is charged with advising the Secretary on policy issues related to ag biotech and with maintaining an intensive dialogue to explore the issues related to ag biotech. The membership includes scientists with experience in rDNA plants, animals, and microbes. It also includes specialists in ecology, biodiversity, forest science, fisheries, human medicine, and public health and epidemiology. It also includes advocates for small farms, consumers, the biotech industry, the public, and an ethicist. All of its meetings are open to the public.

There is some assurance to be taken, too, from the inherently competitive nature of the US political system, a system based on the separation of powers. Congress creates laws; the Executive Branch enforces them; and the Judiciary deals with challenges to them. The Administrative Procedures Act authorizes the courts to review biotech regulatory agency decisions for decisions that are arbitrary and capricious. The Freedom of Information Act demands disclosure of the agencies' deliberations and decisions.

There are other reasons why US consumers may feel protected by the US regulatory approach. The agencies select scientific experts not on the basis of who they are, or with whom they are affiliated, but on the basis of what they know. While the best available scientists are selected for conducting and reviewing tests, regulatory decision-making is not limited to scientists. Non-scientists provide input through public comment, administrative hearings, and even initiation of laws, regulations, and lawsuits. Courts that find agencies non-compliant with the law can impose civil or criminal penalties.

The US system of regulated environmental releases appears to have worked so far. As noted in the Introduction to this book, more than half of the US soybeans and cotton, and nearly a third of the corn, grown in the summer of 1999 were genetically engineered.[18] Other countries have also approved GM crops for commercial production, including Argentina, China, Canada, and Japan.

But the global case asserts, even if no human health risks have emerged so far, it is equally true that *ag biotech has not produced any products that are clearly in nature's or humanity's interest.*

There seem to be a number of counterexamples, as there are thousands of foods containing GM products currently on the US market, including children's breakfast cereals, such as Kellogg's Corn Flakes. Indeed, most processed foods contain GM products insofar as genetic modification is now used to produce food enzymes, amino acids, peptides, flavors, organic acids, polysaccharides, vitamins.[19]

Recombinant alpha amylase is used in high fructose corn syrup, the sweetener in regular Coke and Pepsi and other popular soft drinks. In 1995, this GM product was granted GRAS status. For diet soft drinks, GM aspartame is used, a recombinant amino acid now found in five thousand products on grocery shelves in the US. Monosodium glutamate (MSG), a popular flavor enhancer, is also produced using GM techniques, as are these GM enzymes: glucoamylase, pullulanase, transferase, maltogenic amylase, and sylanase (all starches); xylanase (used in the baking industry); decarboxylase (brewing industry); and, in the juice processing industry, pectinesterase.

Consider chymosin and the cheese making industry. Cheese is produced using rennet, a product traditionally extracted from the stomachs of veal calves and turned into chymosin. Chymosin in turn reacts with a milk protein, casein. The end-product of the chymosin-casein reaction is curd, the basic substance out of which cheddar, Swiss, and provolone are made. Until 1994, the only practical sources of rennet were animal carcasses; for every ten calf stomachs ground up in salt water, cheese makers could produce a gallon of rennet.

In 1990, the FDA approved a gene-spliced industrial substitute for rennet, bioengineered chymosin and, since 1994, this GM product has supplied half of the world's rennet. Some 70 percent of all US cheese is now produced using this GM product. I do not know of any critics of GM chymosin.

Consider "golden rice," rice genetically engineered to contain higher levels of beta-carotene, the substance our digestive tracts convert into vitamin A. Children whose diets are deficient in vitamin A have poor eyesight, and are at increased risk for blindness and susceptibility to diseases such as measles. The World Health Organization estimates that some 230 million children worldwide fall into this category. According to the United Nations, at least one million children under age five die each year from diseases related to vitamin A deficiency. [20]

Golden rice (GR) was produced by Ingo Potrykus and others at the Swiss Federal Institute of Technology. They combined two genes from daffodils with one from bacteria, and inserted them into rice. Potrykus's work was funded by the Rockefeller Foundation and the European Commmunity Biotech Programme. The International Rice Research Institute in the Phillipines is working to breed the new rice into the rice varieties currently being grown in Asia. One serving of GR supplies ten to thirty percent of a child's daily vitamin A requirement.

Notice that the research institutions developing GR are largely funded with public monies.[21] This fact is significant insofar as it may speak to RAFI's fear that ag biotech is only being used by powerful transnational corporations to enforce a new bioserfdom on poor smallholders. Once the new trait has been stabilized in commercial rice varieties, the Institute will distribute the GMO seeds free of charge to peasant farmers around the globe. Therefore, the rice seems impervious to criticisms that it will harm children and hungry people in developing countries.

However, global critics might respond with the following objections, objections that must be addressed.

> **Objection:** GR is a reductionistic strategy that does not address the *systemic* long-range problem: that rice monocultures years ago displaced mixed farms on which green leafy vegetables, carrots, and other sources of vitamin A were once grown. This displacement has caused the current problem.
>
> **Response:** True, perhaps, for some areas of rice production, but not for all areas. But the relevant point is that GR can address a specific short-range problem. The systemic problem of a need for a more diverse, democratically-owned agriculture, cannot be solved by new technology: only government and economic reform can accomplish this task. It seems unfair to criticize this particular technology for its inability to insure such difficult reforms.

Objection: There are alternatives, such as providing children with vitamin A pills, green leafy vegetables, or carrots.

Response: True. But giving children pill supplements does not help their families learn to provide for themselves, and introducing vegetables into the agricultural rotation system requires extensive economic, political, and agronomic reforms that may take years. GR can begin to deliver benefits shortly after its first growing season.

Objection: GR will not solve the problem because it requires that children eat many bowls of rice a day to get their full dose of vitamin A.

Response: True again. GR will not solve the entire problem by itself. But notice that children already eat many bowls of rice a day in these areas; it is their staple food. So GR would not require extensive changes in their diet, or special on-going coaching of children to take their pills. And GR will solve the problem for those children currently consuming sixty or seventy percent of the daily recommended vitamin intake and needing only the modest boost GR provides.

Objection: GR will harm local subsistence agriculture by encouraging rice monocultures to replace current mixed cropping patterns.

Response: This objection is puzzling in light of the first objection, that rice monocultures long ago replaced indigenous rotational schemes. It seems unfair first to criticize GR rice for not addressing a long-standing agro-economic problem, and then to criticize it because it will introduce that same problem.

The global critics are right on one important point, that GR is not a panacea. It will need to be combined with sources of other vitamins and nutrients in order to provide a balanced diet. GR is, however, a commendable new technology that can assist us in meeting our positive duty to aid the needy.

The team that produced golden rice is also working to engineer higher levels of iron into rice. Anemia caused by iron deficient diets is widespread in developing countries, and has been called "the world's worst nutrition disorder," afflicting some two billion people. Might ag biotech soon help, in a small but real way, to save children in developing countries from the scourges of blindness and anemia?

There are other ag biotech products, some with few or no detractors, such as the drug to combat pseudorabies in swine, the first genetically engineered vaccine to be approved for use by the USDA, allowed on the market in 1986.[22] Veterinary biotech products such as the pseudrabies vaccine brought me right back to my first question: How do I square global opposition to the use of genetic engineering in agriculture with virtual carte blanche approval of its use in pharmaceutical production? I did not, and do not, for example, know of critics of a GM test to find prostate cancer. Approved by the FDA in 1987, prostate-specific antigen (PSA) has enabled early successful detection of many tumors during the last dozen years. The PSA test scores well by measures of social justice, too, as prostate cancer affects a disproportionately large number of African American men.[23]

There are now GM drugs not only to diagnosis but to treat cancer, and GM medical tests and vaccines to diagnose and treat Hepatitis B.[24] In patients with multiple sclerosis, a potentially fatal disease, the myelin sheath surrounding the nerves is compromised, leading to dizziness, nausea, and an inability to control one's body. In 1996, Biogen began marketing Avonex, a genefactured form of interferon beta, and a treatment now used successfully by more than fifty thousand people to slow the advance of the symptoms of MS. Gene hunters have shown that single genes are responsible for causing at least two diseases: cystic fibrosis and Duchenne muscular dystrophy. With the genes identified, it may be possible for researchers to devise strategies to overcome the genetic deficiency.

Nor do I know of opponents of the FDA's 1988 approval of products targeted at diagnosing acute coronary syndrome, breast cancer, and Crohn's disease, and preventing Lyme disease. Since the late 1980s, microorganisms have been grown in quantities large enough to produce cheaper and purer supplies of drugs to treat dwarfism, diabetes, arthritis, and forms of cancer other than prostate and breast cancer. Scientists and physicians readily admit that these products are not perfect; the use of interferon to treat cancer, for example, is marred by serious side-effects. Yet many of the so-called miracle drugs have saved lives.

Set aside, if you will, the argument that insofar as we accept medical biotech we should accept ag biotech. Ignore the argument that many publicly-funded scientists are involved in efforts to produce more diverse, nutritive, productive, blight-resistant and efficient strains of staple crops like rice, cassava, potatoes, grains, and tubers. There is yet a strong argument to reject **(1),** an argument that does not depend on either of these arguments. It takes its bearings from Stich's incoherence argument.

Grant the premise of the critics that ag biotech may not solve world hunger. It is equally true that *not* having ag biotech may also not help solve world hunger. It does not seem fair to fault those defending the active development of ag biotech for not being able to guarantee result *x,* when it is equally true that those defending the active non-development of ag biotech are similarly not able to guarantee result *x*.

These were the arguments that finally convinced me to reject **(1).**

(2) There is a substantial risk that ag biotech will do more harm than good to humans, by perpetuating social inequities in developed economies where it will lead to advantages for larger agribusiness farmers that will be unjustly denied to smaller *family farmers*.

Can medium-sized family farms be saved? At the beginning of this chapter, I explained why I have given up hope of a positive answer. Ought implies can. If there is no way to save family farms, then we cannot have a moral obligation to save them. In the conclusion to my earlier edited volume, I suggested that we simply assume that there is a good way to save family farms and try to figure out what it is. Unfortunately, it appears that there is no way to use new technologies to insure that medium-sized farms will be able to compete with small hobby farms and large industrial farms. If we cannot save family farms, then we cannot have an obligation to save them.

I do not oppose mixed, small- and medium-sized, owner-operated farms. I hope, and expect, that many such farmers will find innovative ways to stay in business. I hope consumers will patronize them. I oppose only the use of public monies for commodity supports that aid primarily our largest farmers when those monies might be used in other ways to assist society's truly worst-off.

(3) There is a substantial risk that ag biotech will do more harm than good to *subsistence farmers*, **by perpetuating social and economic inequities between developed economies with their well capitalized farmers, and developing economies, with their under capitalized farmers.**

We have previously noted the fact that world markets for ten major Third World exports are shared monopolies controlled by a half dozen multinational corporations (ch. 2). This a matter for concern because the more concentrated an industry is, the more the companies in that industry can control, and inflate, prices. But in ch. 2 we did not ask the next question: What follows from this fact for ag biotech?

First, banning ag biotech will not fix the problem. Even in the absence of ag biotech, the food industry will still be a concentrated industry with just a few multinationals controlling the markets. Second, singling ag biotech companies out for special censure seems unfair in face of the fact that nearly half of all industries in the U. S. are also oligopolies.[25] It also seems like a step directed at the wrong target. For if the problem is oligopoly power, we ought to address that problem systematically, not by targeting biotech but by addressing every highly concentrated industry sector, such as the computer industry, photo equipment, refrigerators, aircraft, tires, and cars. Third, allowing ag biotech to proceed could allow less developed countries to develop niche-market products, a potential boost to their economies.

The central issue may come down to a matter of choice. Some would prevent developing countries from using biotech, others would encourage it. But who should decide the matter? Relatively well-off and well-fed defenders of the environment in developed countries, or relatively poor, ill-fed farmers in the developing countries?

Michael N. Kibue is a Kikuyu farmer, an African who grows a variety of crops and livestock on a two acre plot on the slopes of Mount Kenya. Kibue, like many farmers in his community, is dissatisfied with the amounts of chemicals necessary to grow cash crops such as coffee and tea, and he is part of a Kikuyu movement to restore traditional farming practices in Kenya. Nonetheless, he perceives an important role for ag biotech. He envisions biotech assisting in the development of more efficient fermentation procedures, so that farmers can diversify the crops they grow for their families. He sees "an urgent need to develop appropriate biotechnology tools and equipment for use at the small-scale farm level," including fermentation vessels, stirring and separation techniques, and food preservation processes (e.g., packaging plants).[26]

Those opposing ag biotech must reckon with Kibue's position. One of the greatest dangers to the well-being of Southern peasants is the fact that for most of the year, fruits and vegetables are not in season. In order to have access to a steady, safe, supply of fruits and vegetables, therefore, farmers must preserve them. The processes for safe canning are not always well understood, and decayed or tainted foods threaten serious harms, including botulism and salmonella. Refrigeration is one answer, but the amount of energy required for refrigeration is not always available in the world's poorest regions. Biotechnology offers the hope of providing drying, preservation, and fermentation processes that would expand the length of time fruits and vegetables were available to the world's neediest consumers.

Kibue is not an ag biotech enthusiast. He argues most strongly for a recovery of traditional patterns of farming and for the reinstitution of indigenous agricultural knowledge. But he recognizes that the revival of these practices, along with a recovery of the use of herbs in traditional medicine, might well benefit from biotechnology. The new biotechniques, he believes, will "help our traditional doctors develop better preparations and prescription methods." Small-scale African farmers, Kibue concludes, "will have to take risks and innovate. Only then do they stand a chance of bettering their lives."[27]

Economic development implies a growing economy. Agricultural history suggests that, in the US, an expanding off-farm economy has provided jobs for farm children for whom there would have been little future on the farm. The problem, of course, is that developing countries often cannot compete with developed economies because their natural resources, infrastructure and education are not as high, putting them at a competitive disadvantage. Displaced farmers in these countries may have a difficult time finding new jobs and recovering their sense of independence and dignity. The slums of Mexico City, Sao Paola and Calcutta, and the so-called "bean riots" of Brazil in the late 1970s, are examples of the effects on real people of the vicissitudes of the world food market. But is the only answer to this problem to try to keep the economy an agrarian one?

Martha Crouch has argued that poor people in developing countries ought to be able to grow some of their own food, without having to compete in the international market economy. Kibue would not seem to disagree with her on this point. But he seems to disagree with an implication of Crouch's position, that ag biotech threatens the ability of Southern peasants to grow subsistence crops. Assume that we take as a goal that we want to shield peasant farmers from the harmful effects of a competitive international market economy. Does it follow that ag biotech worldwide must be shut down?

It seems that the issue comes down to a question of autonomy. Who should decide whether small peasant Southern farmers have access to GM crops? Policy-makers in developed countries? Philosophers in their armchairs? Governmental authorities? Transnational corporations? Or the farmers themselves? Banning ag biotech would clearly take the decision out of the hands of farmers such as Kibue.

(3.1) There is a substantial risk that ag biotech will do more harm than good to *women and children* in developing countries.

In ch. 2 we noted serious questions raised independently by Crouch and Lipton concerning possible ill effects of MA on the groups named in **(3.1)**. And yet the answer of shutting ag biotech down does not seem to address the problem. Compare ag biotech to the green revolution. Neither Crouch's nor Lipton's questions call into doubt that claim that the green revolution's high-yielding seed varieties were in the best interests of people in developing countries. Rather, they call into question the integrity of the political institutions and cultural practices of some of the countries in which the green revolution occurred. Assuming what seems to be true, that the political spheres of some developing countries are corrupt, then the problem is not with ag biotech but with the infrastructures of the countries in question.

Ag biotech may be part of the solution if the countries themselves can innovate, develop their own comparative advantages, and increase incomes. Partha Dasgupta, an economist at Cambridge University, has shown that increasing cash income in a developing country is not always correlated with increasing inequities in power, nor in environmental degradation. Dasgupta cites World Bank research on sub-Saharan Africa suggesting positive correlations among poverty, fertility, and environmental degradation. He cites with approval the conclusion reached by the United Nations (UN) Conference held in Cairo in 1994. The UN argued that poverty in developing countries is best addressed by protecting women's reproductive rights. By empowering women and raising the overall standard of living in poor rural areas, women have the freedom and incentive to limit the number of children they bear.[28]

As Dasgupta suggests, it is very difficult to make valid blanket statements about the impact of MA on developing countries for, as Pierre Crosson also notes, different regions had very different rates of growth. Overall, from the period 1951-55 to 1978-82, developing countries increased food production by an average of 0.5 percent annually on a per capita basis. But East Asia and Latin America did much better than average, at 1.4 percent and 0.9 percent, while Africa did much worse, with food production actually decreasing 0.5 percent.

Crosson believes it is difficult to say whether the green revolution was equitable intragenerationally because the evidence is inconclusive. He thinks the evidence suggests, however, that small farmers and landless workers typically "shared in the higher income yielded by the new technology," because demand for family and hired labor tended to accompany use of the green revolution varieties.

The verdict on the green revolution is mixed, depending on the geographical area selected. For example, in the Indian Punjab, improved varieties of wheat, rice, and potatoes doubled farm incomes within a decade of being introduced. Farmers were encouraged to become innovators in their operations as cash rent policies replaced the previous owner-tenant system. The non-farm economy also grew, bolstered by an adequate education system and a skilled work force. Displaced farmers were able to find jobs in a region where per capita income "has increased 3.0 - 3.5 percent annually for the last two decades."[29]

Yet, the Punjab is not representative of all developing countries. Its population is not as dense as other parts of Asia, a distinct advantage when it comes to changing land ownership patterns. Furthermore, it had already instituted irrigation when the green revolution began. Nonetheless, we have here a region where the benefits of MA "were widely dispersed among people . . ."[30]

Elsewhere, however, the outcome of the green revolution may not have been equitable, "because successful adoption . . . [of MVs] generally requires irrigation, which most farmers in the developing countries do not have access to." As Crosson notes, farmers in these other areas may actually have "suffered diminished income to the extent that the green revolution reduced prices of the crops they produce."

That said, the overall picture may on balance be hopeful because the availability of water is not currently a limiting factor in most developing countries. According to the Food and Agriculture Organization, there is sufficient water available for developing countries to increase by a third the amount of land currently in production. Latin America is currently farming only about 15 percent of its arable land; Africa is farming only 25 percent of the acres it could farm.[31] The limiting factors are the countries' abilities to develop appropriate technologies and the infrastructures required to "assure equitable distribution of the resulting income." According to Crosson:

> Land tenure systems which concentrate most land in the hands of a few farmers or which put in question the year-to-year tenancy agreements, or which put rentals on a crop share basis, may weaken incentives of both large and small farmers, and of both owners and renters to invest in new technology. Large landowners . . . [are likely to be induced] to adopt land-using, labour-saving technologies appropriate to their own resource position but inappropriate where, for society as a whole, labour is abundant and land scarce.

Crosson concludes that "experience with the green revolution strongly suggests that technologies can be developed which meet both productivity and equity criteria."

The lesson seems to be that developing countries need just institutions, honest politicians, land reform, loan forgiveness, and well-developed educational systems so that peasants can act on more complete knowledge, make informed choices, and trust that their institutions will be transparent and their decisions will be honored. But they probably do not need others to make choices for them about which technologies to use. As another Indian writer, R. S. Swaminathan, winner of the World Food Prize, writes, agriculturalists in the developing countries,

> must not worship any tool but should use such combinations of tools and techniques which can help us to reach our goals

> speedily, economically and surely. In other words, we need a blending of what are called traditional and frontier technologies. . . . Land is shrinking and biotic and abiotic stresses are increasing. . . . Disparities between the rich and the poor are growing in every sphere of life. Biotechnology offers scope for adding a dimension of resource neutrality to scale neutrality in technology development. Let us take advantage of this opportunity.[32]

(3.2) There is a substantial risk that terminator technology will do more harm than good to *farmers who save seed* in developing countries.

Terminator technology is a process engineered into crops by multinational corporations intent on protecting intellectual property rights (IPR). The terminator gene insures that the crops will flourish for only one year, and then be rendered sterile. The technology effectively prevents farmers from saving their seed and using it from year to year. This clearly is a harm to them. But will the multinational corporations develop region-specific crop varieties if they cannot profit from doing so? Not to have the assistance of major research and development efforts by corporations might also harm these farmers.

But suppose terminator technology is an unmitigated harm to indigenous seed-savers. Is it the only possible application of ag biotech for the developing world?

Work to produce perennial food crops that will reproduce asexually is underway at the USDA. A process known as apomixis already allows many plants, including crab apple and citrus trees, lawn grasses, and blackberries to produce exact genetic copies of themselves, year after year. The world's major food grains--corn, wheat, and rice--all reproduce sexually. However, if apomixis can be achieved in these crops, farmers would only have to buy seeds once. Thereafter, crops would essentially produce clones of themselves, providing continuous annual yields without the cost of buying new seed.

Here is an example of a potential benefit of ag biotech to farmers saving seed in developing countries. However, as Andrew Pollack has pointed out, academic scientists working in the field are worried that if a large seed company discovers the technology first and patents it, that access to the technology could be denied to Southern subsistence farmers. Concerned about this possibility, an international meeting of researchers in

Bellagio, Italy, issued a document that called for “broad and equitable access to plant biotechnologies, especially apomixis technology.”[33]

(3.3) Ag biotech will lead to an increasingly unjust gap between the world’s poorest and richest.

The effects of ag biotech on the distribution of wealth remains a worrisome concern. The gap between the world’s wealthiest twenty-percent and its poorest twenty-percent has widened in past years. This is not a trend we should accept.

But there are many complexities involved in addressing it. The first thing to be said is that it will not do to interpret Rawls as holding that technologies should be rejected if they do not redress the problem. In my article on bGH, I had misinterpreted Rawls’ in just this way. An anonymous reviewer’s report from a journal to which I had submitted a version of “The Case Against bGH,” helped me to reassess my interpretation by challenging my claim that Rawl’s second principle of distributive justice argues against adoption of bGH. The reviewer argued that I had made two mistakes.

The first mistake was that I ignored Rawls’ principle of liberty, which requires that we set up the basic structures of society so that people will be maximally free. Rawls introduces his view with a famous thought experiment, the “veil of ignorance.”

Imagine that you are in a conversation with everyone else in society. Your collective task is to set up the fundamental structures of society—the economic system, the political system, the network of social services and institutions. And you must set it up in such a way that everyone will be satisfied with the structure, even though they do not know ahead of time what social station they will have in that system. Rawls believes the just society will be the society that results from this conversation. He further believes that the outcome of the conversation will be as follows.

Since people will not know ahead of time in what station they will find themselves, everyone will want to pretend as if they do not know their race, age, gender, social or economic class. Coming to the bargaining table with the needs of everyone in mind, the parties will not bias the structures of society toward any particular group. They will want to arrange things so that they would be willing to accept whatever place they received if it were determined by a lottery that randomly assigned some individuals to be poor, some to be rich, and some to be in the middle.

Under these conditions, all bargainers have a strong incentive to agree to arrange the basic political and economic structures according to three principles. The first principle is equality of rights. There would be no slave class to be exploited by others if all of the bargainers faced equal chances of ending up in the class themselves. All people should be allowed the freedom to engage in actions (and invent and adopt new technologies) that will benefit themselves and others as long as so doing does not unduly limit others' liberties.

The second principle is true equality of opportunity. In western society, we have formal equal opportunity, a policy that states that all applicants will be given equal consideration. However, since many people do not apply for opportunities, we do not have true equal opportunity. Once the prejudicial effects of sexism and racism have been removed from society, then true equality of opportunity can occur.

The third principle is the difference principle: Whenever a society's goods are unequally distributed, any changes in society must favor those who are worst off. The principle entails that no changes in the political or economic sphere shall be permitted if the change harms, or forecloses opportunities otherwise open to, those at the bottom. One of the implications of the difference principle follows. Suppose we have a choice between a public policy (*p*) that will make the poor a little bit better-off, or public policy (*r*) that will make the rich a little better-off. If we can only choose to enact one of these policies, we ought always to choose (*p*).

The principle of equality and liberty implies that technological innovations are to be highly valued in this society, all other things being equal, insofar as they are the result of people exchanging ideas and discoveries as free equals. To ban a new technology would be *prima facie* acting contrary to the principle of liberty unless the new technology will clearly limit the liberty of others. I argued in ch. 1 that bGH would unfairly limit the liberties of family farmers. Did it?

As we have seen, widespread adoption of bGH may have restricted the range of some farmers' choices by foreclosing the choice of continuing to run a dairy farm, but it did not change the society in which these farmers' other choices were protected. Nor did it foreclose their ability to choose to convert their operation from dairy to, say, a cash grain farm. Nor did it foreclose their ability to receive compensation upon the sale of their property. So bGH does not seem to run afoul of any of Rawls' principles. It seems, in sum, that one cannot justify banning bGH by appealing to Rawls' social contract theory. That was my first mistake.

The second was that I misapplied the difference principle. The way in which Rawls' idea of justice protects those at the bottom of the ladder is not by banning specific technologies but rather by banning changes to the fundamental structures of political and economic life that would discriminate against the poor. As the journal reviewer pointed out, my argument against bGH not only ignored Rawls' liberty principle, but misinterpreted the difference principle "as applicable to particular policy areas." Rawls himself, wrote the reviewer, "repeatedly asserts that this is completely unjustified." [34]

Distributive justice requires that we set up institutions so as to promote fairness and improve the lot of the worst-off. Will ag biotech exacerbate inequalities in distribution of wealth? Bryan and Farrell argue that the benefits of new technologies will in the long-term reach those at the bottom of the world's poorest countries.[35] To the contrary, global critics of ag biotech cite figures suggesting that precisely the opposite effect will occur. In 1960, the ratio of distribution of income between the twenty percent of the world's population that lived in the richest countries compared to the twenty percent that lived in the poorest countries was thirty to one. Today that ratio has widened, doubling, in fact, to sixty to one.[36] Faced with those figures, one might conclude that ag biotech is certain to increase injustice in the world.

But is such a conclusion inevitable? Perhaps not. First, one might argue that new progressive policies could effectively prevent technological developments from increasing global inequalities. In the case of bioprospecting, for example, we might be able to institute effective international regulations requiring standardized profit-sharing contracts between international corporations and the indigenous groups that provide biologicals.

Second, one might argue that the statistics themselves do not prove injustice. Granted, globalization has been accompanied by a troubling increase in relative income inequality. However, what would have happened in the absence of globalization? We do not know the answer. It is conceivable, however, that many of those at the bottom might be even worse-off today than they would have been in the absence of the processes of globalization. If so, then globalization would measure favorably by Rawls' standard of justice.

That said, we must discuss the realities of technology transfer. The multinational companies that have developed the first wave of ag biotechnologies have protected their intellectual property with patents. In order to use these new technologies legally, poorer countries must pay the fees that these companies charge. Unfortunately, the poorer countries lack the financial resources needed to pay the rents, and lack the human resources needed to negotiate the legal intricacies of intellectual property.

If a country wants to plant vitamin A rice, it must first complete a complex process of negotiation, even though the rice was developed by a public agency, IRRI, that wants to provide the rice seed free to the country's farmers. The reason is that the basic genetic knowledge and accompanying technologies necessary to produce the rice are owned by companies such as Monsanto and Du Pont. As C. S. Prakash explains,

> If Vietnam or Liberia wants to distribute golden rice seeds to its farmers, it must first negotiate with various companies for the gene transfer, gene promoters and selectable marker technologies that were used in its development. . . . Thus, agricultural biotechnology cannot make inroads into developing countries without a 'freedom to operate' license from the owners of these technologies--major life science corporations.[37]

Prakash concludes that "industry 'ownership' of genes and technologies used in the development of such varieties represents a serious obstacle," adding that

> If companies really want to help combat global poverty and hunger, they must make their technology available for use by developing country farmers on select food crops such as rice, cassava and millets on a royalty-free basis.

Premise **(3.3)** clearly identifies an area of concern. Given the political obstacles that stand in the way of realizing ag biotech's promise to feed the world's hungry, one might be tempted to defend the following claim.

(3.4) We should allow publicly-, but not privately-, funded ag biotech research and products.

One might be opposed to allowing the profit motive to enter the sphere of genetic manipulation of the environment but in favor of allowing genetic engineering for the development of crops in poorer countries. What would be wrong with supporting the production of biotechnologies like vitamin A rice with public dollars, and requiring the free distribution of the technology, while banning the development of ag biotech in the private sector?[38] This strategy would allow for a more deliberate development of the technology, free of the urgency that accompanies profit-driven technology development.

It is an interesting argument, but it has two flaws. First, there is the problem of fairness. Assuming that both the public technology (e.g., vitamin A rice) is safe for consumers and the environment, and that the private technology (e.g., chymosin) is similarly safe for consumers and the environment, why should public agencies be given a freedom denied to private companies?

There is a deeper problem. The development of publicly funded ag biotech may not be possible without the assistance, financial and intellectual, of the private sector. Monsanto helped to develop vitamin A rice. It is an open question whether the rice could have reached its current state without the assistance of Monsanto. If private companies are contributing essential knowledge and monetary resources to public efforts, then the public efforts, by definition, would not be possible without the accompanying commercialization of products for which wealthier consumers are willing to pay. Therefore, **(3.4)** seems to be neither fair nor practical.

(4) There is a substantial risk that ag biotech will do more harm than good to *scientists and consumers,* **because scientists must increasingly pursue reductionistic, short term, applied science to benefit private corporations, rather than holistic, long term, theoretical science to benefit public taxpayers.**

The problem with this objection is that it suggests that there is a single dichotomy between sciences, reductionistic versus holistic, and that the first is bad and the second good. In fact, there are many different kinds of science, and the goodness or badness of practicing a science is assessed by whether the science is practiced according to standards internal to it. Consider nuclear physics, inorganic chemistry, and molecular biology. These sciences are by their very nature reductionistic, and progress is made in them only through very compartmentalized linear thinking and by repeating experiments to confirm observations.

Other sciences proceed differently. The so-called historical sciences, geology, much of evolutionary biology and astronomy, study objects that by their nature are not subject to controlled repeatable experiments. Therefore, geologists and astronomers must engage in more holistic theorizing in order to explain their subjects.

It seems that critics of ag biotech are largely correct: the funds currently available to molecular biologists in agriculture overshadow the funds available to traditional plant breeders, agroecologists, and environmental scientists. But this situation may change as more and more genes are identified and the research establishment begins to try to move these genes into established crop lines. The current funding situation need not suggest either that molecular biology will forever be favored over other biological fields, nor that reductionism as a method threatens to overtake all of the sciences.

In chapter 2, I argued that scientists need to take a more holistic approach. But such an approach is not incompatible with reductionistic approaches. For reductionistic scientists such as molecular biologists to succeed in transferring their results to agriculture, they must partner with traditional plant breeders. For holistic scientists such as traditional plant breeder to have access to all of the potential sources of new traits and varieties, they must partner with molecular biologists.

(5) There is a substantial risk that ag biotech will do more harm than good to *future generations* by foreclosing possibilities for them to feed themselves.

This objection is a variant on the objection discussed at the beginning of this chapter, where we observed an incoherence in the precautionary principle. The Stich argument defeats **(5)**. For while it may be true that the development of ag biotech will endanger the future food supply, it is equally true that the failure to develop ag biotech may similarly endanger the food supply.

But formulating our options as strict either/or dichotomies does not allow for compromise solutions. Can we not pursue ag biotech in the context of pursuing sustainable agriculture? A sustainable agricultural system would be compatible with the environment, and a sustainable system with ag biotech in it might make it easier for future generations to feed themselves.

What is a sustainable agricultural system? The answer is hotly contested, but Pierre Crosson has probably done as much as anyone to work out a way to balance conflicts over the rights of present and future human generations with respect to the environment. A sustainable system of food production, he argues, must satisfy "demands for food into the indefinite future while meeting equity conditions in food production both within and across generations."[39] Notice that this definition does not argue for preserving farmland because of its inherent worth. Crosson argues that it ought to be preserved because it will be needed by future generations of humans. But the issue of the inherent value of nature aside, how are we to determine among us humans *whose* needs should receive top priority?

The answer of the liberal tradition to this is clear: the poor. Rawls has argued most forcefully for the appropriateness of our intuition that those at the bottom of the economic and social ladder ought to have their needs considered first. Crosson's definition of "an equitable, therefore sustainable" agriculture is consistent with Rawls' view:

> I define an equitable, therefore sustainable, system of food production as one which indefinitely meets rising demands for food and fibre without incurring rising economic or environmental costs and which distributes income in a way regarded as equitable by the least advantaged participants in the system.[40]

Crosson's definition is attractive because it incorporates criteria of both intragenerational and intergenerational equity. The criterion of intergenerational equity is that each generation is responsible "to manage its agricultural resources so as to pass an unimpaired capacity to produce food and fibre to the next generation." The intragenerational equity principle is that the food production system

> must yield significantly rising real income for the poor involved in agricultural production. A `significant' increase in income is one sufficient to satisfy the poor that their condition is improving and will continue to improve at a rate such that when they look back every five years or so they will feel that they are distinctly better off in the material things of life.

By insisting that the disadvantaged be given the right to decide whether they are better or worse-off, the definition meets our condition that those at the bottom of the socio-political economic spectrum be allowed to speak for themselves.

Crosson's definition of sustainability does not insure increasing equality in income distribution. "Indeed," notes Crosson, "it is consistent with increasing inequality."

> The condition for equity is that the food production system generate rising real income for agriculturalists and assure that the poor share `significantly' in the gains. It does not necessarily require that they increase or even maintain their relative share of total income. As defined, therefore, intragenerational equity is in the eyes of the beholders, where `the beholders' are the poor involved in agricultural production.[41]

Crosson goes on to note that the poor may set "a less exacting standard" if they judge the performance of the system by how their present income compares with their past rather than if they compare their income to the income of others. But they need not assume a lower standard of comparison; the definition allows them to choose their own benchmark.

Crosson goes on to note that while his definition incorporates the environmentalist concern for "non-rising costs" to nature, "the grounds for insisting [on this] is the precept of intergenerational equity, not a commitment to maintaining the integrity of the natural system *per se*."[42] Crosson is concerned about nature, but not because he believes it is valuable in itself but because it is needed as a resource for future generations. Even though we do not now put a price on environmental degradation, he believes, we should do learn to do so. We must learn how to figure costs such as

> erosion damage to rivers, lakes, reservoirs and harbours; effects of fertilizers carried by eroded soil and runoff in stimulating eutrophication of water bodies; human illnesses, deaths and damage to ecological systems from use of pesticides; loss of valuable plant and animal habitat through deforestation and drainage of wetlands; increasing soil and water salinity associated with irrigation, and so on.[43]

And why should we be concerned? Wendell Berry has put it succinctly: "And so the land is taxed to subsidize an `affluence' that consists, in reality, of health and goods stolen from the unborn."[44] Crosson reasons in more economic terms: the external environmental costs "are no less real than the economic costs of production" and yet "those bearing the costs have no way of exacting payment from those who impose them."

> For example, those who suffer loss of recreational facilities because of sediment-laden waters have no way of collecting compensation from the farmers whose fields are the source of the damaging sediment.[45]

We do not have very good data about how costly these environmental externalities have really been, Crosson confesses. He acknowledges that our increasing use of fertilizers has given rise to nitrate concentrations in rivers in parts of the midwest and California that "quite often exceed the Public Health Service standard of 10 ppm.[46] Another estimate put the environmental costs of pesticides at more than $800 million each year.[47] But Crosson thinks it unlikely that these environmental costs would offset the gains made in the area of intergenerational equity. He gives several reasons for this judgment: Since the Second World War, the real income of farmers and farmworkers has more than doubled. From 1960 to 1980, the biggest percentage gains have been made by the smallest farmers, those with total gross annual sales of under $2500. If we suppose that these are the least advantaged people in the agricultural industry, then the criterion of intragenerational equity has been satisfied.

Is MA in the United States sustainable by this definition? Crosson notes three important facts. First, a quarter to a third of all US food production is exported; this food could be retained for domestic use if needed. Second, current rates of soil erosion, if continued for another century, would reduce crop yields by only 5 percent. Third, the rate of land conversion from agriculture to other uses is minimal, around 300,000 hectares per year. Crosson concludes that the current system is sustainable, assuming that export demand does not increase dramatically.

What steps can help to insure that Crosson's judgment is correct? He makes three suggestions. First, we should develop technologies capable of increasing yields on smaller parcels of land, while also "satisfying both cost and equity criteria for sustainability." Because fossil fuels are expected to become increasingly expensive, these technologies should not be energy-intensive. Examples of such technologies would be "improved photosynthetic efficiency and biological fixation of nitrogen by corn plants." Second, figure out exactly what the environmental effects of herbicide use are with an eye on developing alternative means of weed control. Third, insist on more efficient use of ground and surface water for irrigation. All three suggestions represent goals ag biotech could, in the long term, assist in achieving.

Crosson takes all of this as evidence that the modern agricultural system has served the interests of equity. Duvick also argues that ag biotech should eventually contribute to these goals of sustainable ag: crops able to flourish without heavy chemical inputs; diverse and alternative crops for multi-cropping rotation practices; varieties with enhanced resistance to diseases and insects, and crops adapted to various climates and soils, such as cool and wet, or hot and dry, conditions.[48] While the genetic linkage maps and techniques of genetic transformation necessary to accomplish these ends is not yet in place, they are likely to be in the distant future. In that sense, ag biotech is compatible with sustainable ag.

Interestingly, at least one global critic of ag biotech seems to have changed his mind about the compatibility of ag biotech and a more sustainable agriculture. In conversation, Wes Jackson has said that he now believes the techniques of genetic engineering might have one, very limited, purpose in his attempt to build a perennial polyculture for high plains farming: helping to develop a cone to hold the seeds of the grassy species around which he is developing his system.

I have come to believe that I did not give sufficient credit in chapter 2 to people like Homer LeBaron and Mary Potter. We do indeed need a context and sense of balance in our approach. We need to be more skeptical about claims that organic foods (e.g., peanuts) are inherently safer than others, and we need to be more open to the possibility that GM foods (e.g., golden rice) might be beneficial.

I faced an uphill struggle in holding to my global argument that GM foods would harm future generations. But what about the risks to the environment?

(6) There is a substantial risk that ag biotech will do more harm than good to ecosystems, by leading to environmental catastrophe through release into the wild of virulent *genetically modified organisms, plants, or fish.*

The global case asks us to consider the pressures we have put on the land itself. Let us distinguish at least four different claims here, and examine several variants of some of these claims.

> **(6.1) We should not have any ag biotechnologies because** *genetic engineering of any sort is a risk to the environment.*

This claim seems false for two reasons. First, much genetic engineering is confined to the lab. Ban plant and animal biotech if you will, and all genetic engineering of any microbe with the potential to escape into the wild. There will remain some forms of GM food processing that uses microbes to produce enzymes that may be safely quarantined, along with their industrial by-products, in contained vats and closed production systems. To the extent that the existence of such GMOs depend for their very being on the favorable environments of climate controlled lab conditions, these GMOs could not survive outside the lab. By hypothesis, therefore, the processes of these food biotechnologies pose no risk to the environment.

Second, there are examples of biotech serving environmental conservation goals. The bark of the Pacific Yew tree contains taxol which has cancer-fighting properties. After this fact was discovered, the monetary value of the Yew tree rose, and so did the number of them felled. However, the active ingredient of the Yew bark has now been introduced into tissue culture lines, and these lines are producing taxol. The reclamation of degraded soils on military bases and refineries may also be aided by genetic engineering. Modified plants and microbes can be used to concentrate metals, extracting these environmental bad actors into the plant tissues. Taxol and contaminant cleanup seem to be distinct advantages of ag biotech to the environment. So **(6.1)** fails.

> **(6.2) It is impossible to know whether GMOs might** *persist in the environment* **years after release; therefore, it is impermissible to use them.**

One cannot intelligibly deny that there is a small risk of ag biotech producing a product that could bring catastrophic environmental damage. However, **(6.2)** proves too much. For we can be virtually certain that some forms of GMOs will not persist in the environment years after their release. An example is Bt corn, which, like traditional, high yielding varieties, loses vigor after one year. It is, strictly speaking, impossible to know whether a given hybrid corn will persist in the environment years after release. But we do not, on this account alone and in the absence of other considerations, think it is impermissible to plant such corn. Similarly, it is, strictly speaking, impossible to know whether a given *non*-GM corn will persist in the environment years after release. But, by parity of reasoning, we should not, on this account alone and in the absence of other considerations, think it is impermissible to plant such corn.

It will be useful to remind ourselves here of the environmental unfitness both of MA's varieties and of GMOs. As Brill argued in ch. 2, new plants must have a variety of favorable traits in order to outcompete other plants: their seed must be able to be dispersed widely; the seed must be able to survive a long time; the plant must grow more quickly than others around it; and so on. Current GM varieties do not seem to have these properties, and we could, if we desired, require that all of them contain terminator genes, so that they could not persist in the environment. So **(6.2)** seems unpersuasive.

> **(6.3)** *Every GMO persists in the environment* **years after its release; therefore, it is impermissible to use them.**

(6.3) is false. First, some GMOs are born, thrive, and die in industrial vats, never being released into the environment. Second, of GMOs that are released into the environment, many GMOs do not persist for years. In 1995, scientists in the United Kingdom released an innocuous free-living bacteria into a wheat field.[49] They found that it spread more rapidly than they had been led to expect from experimental greenhouse studies, probably because there was more rain than usual and water containing the bacteria percolated through the soil. Fifty days after spraying a row of plants, the bacterium could be found on unsprayed plants in rows next to the treated plants. While the GMO had a greater ability than expected to disperse, it nonetheless had limited ability to survive. Less than a year after the application, scientists were unable to detect the GMO anywhere in the plot.

> **(6.4)** *Even if we cannot detect GMOs persisting in the environment* **years after release, it is too dangerous to use them.**

(6.4) is too vague to be of any use. We can study risk factors involved in deliberate release, such as the effects of gene transfer on: target plant growth and vigor; nontarget plants and animals; ecology; dispersal through soil and water; changes in pathogenicity and host range; creation of new pathogens; etc. And we can design statistical experiments to give us data about the relative percentage of change involved in releasing traditionally modified and genetically modified organisms. And these statistical results can give us an idea of which sorts of changes are within the standard range, and which sorts fall outside it. And we can then debate how great a variation in results we are willing to accept as "safe." But **(6.4)** does not tell us how to make such determinations. It declares universally that GMOs are always dangerous, without explaining why. We can be virtually certain that Pioneer's hybrid corn will not persist in the environment years after its release, because the seed dramatically loses vigor after one year. We cannot detect hybrid corn persisting in the environment years after its release. If we accept **(6.4)**, we should deem hybrid seed corn too dangerous to use because we cannot detect hybrid seed corn persisting in the environment years after release. By substituting "seed corn" for "GMOs" in **(6.4)** we get a counterintuitive result.

> **(6.5) Not knowing what the risks are,** *the public makes decisions in an uninformed way*; **therefore we should not use GM crops.**

The conclusion of **(6.5)** does not follow from its premises. The fact that the public is largely ignorant of the risks of doing *x* does not justify banning *x*. Indeed, nothing seems to follow from the fact that the public is ignorant about something. The public is largely ignorant of the risks of driving cars painted red, or of eating shellfish. The public is also ignorant of the risks of *not* proceeding with ag biotech. Does it follow that we should ban the driving of red cars and the eating of shellfish?

> **(6.6) There is a substantial risk that ag biotech will do more harm than good to ecosystems, by leading to environmental catastrophe through release into the wild of virulent genetically modified *plants*.**

Turning our attention from organisms to plants, there have, of course, been cases where ecosystems were overrun with "exotics," new plants introduced from other areas. Examples include the kudzu that chokes Southern trees and the hydrilla that clogs Southern waterways.[50] But these are not cases of mutant plants spreading a new genetic inheritance throughout the ecosystem. They are imported plants brought into ecosystems where no natural checks were in place to limit their growth.

It will help to remind ourselves here of the problem of gene flow into other varieties is a valid concern. Monsanto's web-page press releases claims that "weed resistance to Roundup is much less likely to occur than resistance to most other herbicides . . . because Roundup herbicide possesses unique traits." The unique trait is the mode of action of Roundup. Roundup "inhibits EPSP synthase." Furthermore, Roundup is a post-emergent herbicide with no residual soil activity — greatly limiting the chance that resistant weeds over time could appear in a weed population."[51]

But Monsanto's claims notwithstanding, there are specific cases that need attention. Danish researchers have shown, for example, that genetically engineered herbicide resistant canola is a highly outcrossing plant that can pass its herbicide resistant genes to weedy relatives.[52] The GM canola also tends to volunteer readily. This situation is potentially worrisome because if the gene for herbicide tolerance is passed to the weed, then the herbicide will no longer be able to kill the weeds.

The case deserves a closer look. The first Roundup Ready canola crop was field tested in 1991. In 1995, the major regulatory agency in Canda, Agriculture and Agrifood Canada, examined the case carefully. Ecological risk assessment proceeds by examining two factors. "Exposure" denotes the probability of a harmful event. "Hazard" denotes the degree of harm involved. Some very harmful hazards (e.g., the creation of an eggplant that could eat Chicago) have a very low exposure. Some events carrying a very low level of hazard (e.g., the creation of a localized herbicide resistant weed that dies out after one generation) have a comparatively high level of exposure. Canada was concerned about canola because the GM Monsanto canola, *Brassica napus* var. *oleifera,* easily interbreeds with two abundant weeds, *B. rapa* and *B. juncea*. The risk of exposure, therefore, was assessed as high. They found, however, that the herbicide resistant genes conferred no greater fitness on the weeds. Consequently, "currently accepted weed management measures" were deemed sufficient to control the weeds.[53] While exposure to gene flow in some cases may be high, the hazard may be low.

The problem of gene flow into other varieties is a valid concern. The problem of a GM variety itself becoming a weed seems not to be a concern.

We noted in ch. 2 the concern of potential cross-pollination between GM crops and wild varieties. Margaret Mellon and Jane Rissler, have pointed out that in North America, not only canola, but also carrots, sunflowers, radishes and squash are grown in close proximity to wild relatives.[54] There is reason to be cautious about gene transfer from certain genetically engineered crops crossing through natural means into weeds in or near fields, with the consequence of introducing traits such as herbicide resistance into wild species. The consequence would be the ruin of the genetically engineered technology because the weeds would no longer succumb to the chemicals.

But it is important to be clear about the implications of studies showing that *certain* transgenic plants may cross-breed with wild weed species. Other transgenic plants do not present the same possibility because they have no weedy relatives. There are, for example, no weedy relatives growing near cornfields in Iowa because corn is an imported crop not native to this state. Mellon and Rissler, some of ag biotech's most recognizable critics, assess the risk of transgenes flowing from GM corn and soybeans to weedy relatives as "nil."[55]

The outcrossing problem presents us with a good argument for careful risk assessment and intensive strategic management of GM crops. We need to balance the concerns of the need for increased production, human safety, social usefulness, and environmental compatibility.

Consider two more decisions that seem to attest to the fundamental soundness of the manage-and-regulate strategy. A company intended to market carnations genetically modified for extended vase life or altered petal color. The GMAC, the Australian governmental body responsible for approving commercial releases studied data from small field and demonstration trials and decided that the new flower posed "negligible" risk. Negligible risk was defined as risk that was no greater than the risk associated with the unmodified carnation. "Carnation," writes Kirsty McLean, "has no weedy characteristics and is not closely related to any weed in Australia.

The biology of the carnation is such that there are no realistic ways for the genetically modified plants to escape from cultivation and become established as populations in the wild, or for gene dispersal from the genetically modified carnation to occur. The GMAC approved this release.

The Australian agency did not approve a second release, involving *Bt* cotton plants expressing the CryIA(c) gene. The new cotton carries a protein produced by the CryIA(c) gene that kills the major caterpillars that attack cotton. In justifying its decision, the GMAC wrote that important data was missing on: "1) the consequences of transfer of the Bt gene to native Australian *Gossypium* species, and 2) appropriate resistance management straegies."[56] Instead, GMAC recommended that the release be confined to areas of southern Queensland and New South Wales areas, presumably, where *Gossypium* is not abundant.

These two cases suggest that regulatory agencies are not rubber stamps, and are capable of turning down requests based on environmental and agronomic concerns. Will the system work everywhere? It is impossible to predict what will happen in developing countries with few resources for developing "the system." But it bears pointing out that it is in the interest of developed countries to assist developing countries to establish regulatory mechanisms for the safe use of genetically engineered crops. A program at Michigan State University, funded by USAID and the USDA/APHIS is dedicated to this task. The project is called the Agricultural Biotechnology Support Project (ABSP), and its goal is to help developing countries use and manage biotechnology.[57]

Several countries have gone a long way toward developing the infrastructure needed to insure safety in the use of genefactured seeds, crops, and animals. Egypt has approved national biosafety guidelines, and Indonesia and Kenya have guidelines that are awaiting approval. Representatives of all three countries participated in a workshop, learning biosafety and risk assessment protocols for the handling of GMOs at ABSP.[58]

ABSP is also assisting in strengthening biosafety regulations in Latin America and the Middle East, and in building biocontainment research facilities in Egypt and Indonesia. "However," concludes Andrea Johanson, assistant director of the project, "the lack of institutionalized guidelines and/or field testing regimes, coupled with uncertainty in collaborating country governments, has made the actual transfer of materials difficult."

Prudence and caution are necessary, and we must redouble our efforts to ensure that every country developing ag biotech has adequate safeguards in place and that scientific research results are freely and openly shared. But banning the technology probably will not help these countries.

(6.7) There is a substantial risk that ag biotech will do more harm than good to ecosystems, by leading to environmental catastrophe through release into the wild of virulent genetically modified *fish*.

This is an area of grave concern. Some transgenic fish have been shown to be more fit in some wild environments than their wild counterparts, meaning that the engineered fish may outcompete and perhaps completely replace the wild varieties. Antifreeze polypetide genes have been introduced into Atlantic salmon, tilapia, carp, and giant prawn.[59] The antifreeze gene renders these individuals capable of withstanding colder temperatures than their wild relatives. If these species were to be fish farmed in northern latitudes where they are not presently found, and if a fish or two were to escape from the aquafarm, the exotic species might well colonize large areas, potentially driving other species out of existence. Fish, unlike, say, cattle, can travel great distances and colonize vast areas of ecosystems.

But is this a reason to stop releases altogether? Recognizing the dangers, the USDA released "Guidelines for Research Involving the Planned Introduction into the Environment of Genetically Modified Organisms" in 1992. The Guidelines, developed by the USDA's Agricultural Biotechnology Research Advisory Committee (ABRAC), were necessary to fill gaps in the regulatory environment. As previously noted, various US government agencies regulate GM research, but some research has fallen through the cracks, including transgenic fish. Notably, the ABRAC's Guidelines are not legally binding. Here legislative action is necessary to make adherence binding rather than voluntary.[60]

There are scientific and theoretical considerations that suggest the risk may be acceptable. First, according to Elliot Entis of Aqua Bounty Farms, the GM salmon grow faster, and reach commercial size in two years. Ordinary salmon take three years. But, contrary to some reports about this issue, the GM salmon raised by Entis' farm at least do not grow larger than ordinary salmon. After three years, the GM and non-GM varieties are comparable.[61]

Concerning the ability of GM salmon to outcompete wild salmon, Entis claims that a study by Wayne Knibb suggests that the GM fish are less fit and more likely to be outperformed by wild fish. Entis notes that in one study of the behavior of two groups of fish, the GM salmon "had such a desire to feed that they did not flee from introduced predators" in the way that the non-GM salmon did.

Finally, Entis notes that there is a kind of terminator technology readily available to GM salmon raisers. By administering shocks to salmon eggs after they are fertilized, breeders can induce triploidy, a condition that renders the fish sterile. Entis claims that the success rate for this procedure "approaches 100% in salmon if done properly." Requiring the use of triploidy in all GM salmon could further reduce the environmental risks of unwanted introgression of GM salmon into wild stocks.

Strategies are available for reducing the risk that GM salmon will mate with wild salmon and thereby reduce the wild salmon's ecological fitness. Nonetheless, our history with non-GM farm-raised salmon suggests that the damage done by escaped GM fish could be severe, as super salmon might out-compete wild salmon for food, mates, and habitat. It would seem wise to delay commercialization of this technology until these issues can be effectively addressed.

(7) There is a substantial risk that ag biotech will do more harm than good by ***narrowing plant germplasm*****, and more harm than good to the atmosphere by reducing the quality of** *air, soils, and ground and surface waters.*

This is also an area of legitimate concern. Consider water. Across the country, over half of all water pollution from non-point sources comes from farming.[62] One of the worst affected areas is the Chesapeake Bay region where, researchers believe, the run-off of Furadan applications to protect corn from European borers is responsible for the demise of the Bay's bald eagles.[63]

But how do we decide when water quality has been reduced to a level we should consider morally unjustifiable? How do we decide at what point plant germplasm worldwide has become too narrow? One measure might be: whenever water becomes so degraded, or germplasm so narrow, that *future generations will be unable to feed themselves.* This criterion points us to the issue of the role of ag biotech in sustainable agriculture, our obligations to future generations, and the role of ag biotech. I have previously argued, in section **(5)**, that these concerns, while valid, need not rule out ag biotech.

Another measure would be: whenever *any* water quality or plant variety is lost. This criterion points us to the issue of the intrinsic value of nature and ag biotech's intrinsic disrespect for that value. I have argued in the previous chapter (section **(14)**), that this construal of the value of nature cannot be sustained.

While **(7)** raises important concerns, there are reasons to believe that abandoning ag biotech would not help us address the concerns. Rather than banning ag biotech, one might make receipt of national farm program crop subsidies dependent on compliance with environmental principles of low polluting behavior. Along with policy and educational efforts to encourage non-polluting modes of farming, we could pursue new genefactured bacteria which might biodegrade specific pollutants already in the water.

Finally, the argument must be mentioned that MA has been good for the environment if we assume that humans are part of the environment and have a basic right to be fed. During the twentieth century, the yields of nine major crops have increased from two to sevenfold.[64] Without industrial agriculture, much more land would have to be in production to produce equivalent amounts of food. If yields had remained steady after 1960 instead of continuing to grow, we would need another 10 to 12 million square miles, "roughly the land area of the US, the European Union countries and Brazil combined."[65] There are reasons for thinking that ag biotech may do more to benefit the environment than to harm it.

Let us turn, finally, to the consequences of ag biotech for animals.

(8) There is a substantial risk that ag biotech will do more harm than good to *research animals, livestock and wildlife*, by causing them to suffer or die, or to prevent them from continuing as a species.

Consider wildlife first. Is there a substantial risk that ag biotech will do more harm than good to wildlife?

(8.1) There is a substantial risk that ag biotech will do more harm than good to *wildlife*, by causing them to suffer or die, or to prevent them from continuing as a species.

We noted in some detail in ch. 4 the threat to wildlife habitat from the expansion of MA. It is important to point out, however, the ways in which ag biotech might benefit efforts to conserve wildlife. Currently, wolves on Michigan's Upper Peninsula are failing as the pack becomes inbred. Foreign DNA, taken from wolf packs in Alaska, might be implanted into the eggs of Michigan wolves, thereby improving the genetic biodiversity of the pack and improving its chances for survival. In another example, panda bears are an endangered species; there are only about a thousand of them in the world. Chinese scientists are making progress in their efforts to clone a giant panda, according to the state-run Xinhua News Agency. "Scientists from the government-funded Chinese Academy of Sciences grew the embryo by introducing cells from a dead female panda into the egg cells of a Japanese white rabbit . . . The embryo was nurtured over 10 months, and scientists are now trying to implant it in a host animal's uterus." [66] Animal biotechnology can be a useful tool for wildlife ecologists.

It may also be used to increase wildlife habitat. Consider two possible scenarios. First, if ag biotech increases productivity, marginal land now cultivated might be taken out of production. Farmers in Iowa, for example, might decide to restore the land for use by the wild turkeys and other wildlife that populated the state a century ago.

Second, suppose more of our food were to be "grown" in industrial factories using genetically altered plants and bacteria. More of our food could come from less of our land. The result might again be a positive one, environmentally speaking, as more marginal farmland were taken out of production, easing erosion and chemical run-off.

Of course, there are no guarantees. The pieces of farmland returned to a more "natural" state would, in the absence of public planning, not be connected in any integral way. They would be connected only by the decisions of individual farmers, and almost certainly in a piecemeal and fragmented fashion. Random idling of of selected bits of a farm would not necessarily be good for all wildlife species, even though it might be good for the small game varieties favored by hunters. So, again, public policy would be needed to orchestrate the idling of land so that chunks of land large enough for wildlife habitat could be recreated.

In order to take up the question of transgenic farm animals, it will repay us first to consider animals more generally, and research animals in particular.

(8.2) There is a substantial risk that ag biotech will do more harm than good to *research animals*, by causing them to suffer or die.

It would seem that those with an animal rights theory (AR) must oppose the production of all transgenic animals (TA). Given the radical implications of the animal rights theory that led me to become a vegetarian, it probably sounds odd for me now to suggest that animal rights defenders must endorse the production of many transgenic animals. The incongruity of the claim is magnified by reviewing some of the more troubling experiments which, as recounted in the prior chapter, have resulted in quivering, obese, or even headless TAs.

Contrary to what I wrote in chapter 3, I now believe that the animal rights theory permits, even requires, a certain amount of suffering and death in research animals, although not usually in so-called food animals. The death of a research animal may be permitted in AR if the animal does not have a future and if the situation involves a choice between the loss of an ordinary human life and the loss of the animal.

In what follows, I will argue that AR has two very different interpretations. According to the abolitionist interpretation of animal rights (ARA), all TA research is morally objectionable. However, according to what I will call the reformist interpretation (ARR), much TA research is justified. Finally, I will present reasons for favoring ARR.

When writing chapter 1, I did not understand the full implications of Regan's remark that AR not only permits but sometimes requires the sacrifice of animals to save human lives. In so-called lifeboat cases, in which four humans and a dog are in a lifeboat that can support only four lives total, Regan believes we ought always to sacrifice the dog. Indeed, were there a hundred dogs on the lifeboat, Regan holds that we ought to sacrifice all of them in order to save the four humans. In theory, anyway, no matter how few humans we might save, any number of animals should be sacrificed to save the humans.

> Let the number of dogs be as large as one likes; suppose they number a million; and suppose the lifeboat will support only four survivors. Then the rights view still implies that, special considerations apart, the million dogs should be thrown overboard and the four humans saved.[67]

Regan carefully distinguishes his reasoning from utilitarian reasoning. He points out that the case is a case of having to decide whether to over-ride the rights of the many, or the rights of a few. While everyone has the basic right not to be harmed in order to promote the good of others, there are cases where we are forced to choose between over-riding many people's rights, or a few people's rights. In cases where the harm to be caused to all parties is comparable, Regan avers, we ought to apply the Miniride Principle:

> *Where* **comparable** *harms are involved, override the fewest individuals' rights (MP).*[68]

Comparable harms are equivalent harms. Causing Alice to die is comparable to causing Betty to die, but it is not comparable to causing Betty to have a root canal. *MP* applies only to cases of the first sort, where we must choose to inflict the same harm on one of two individuals. Regan rests the justification of *MP* on the central idea of his theory, the principle of respect, which requires that each individual be treated equally with others.

MP is also applicable to cases involving disparate numbers of victims. If we are forced to choose between causing comparable harms to one or many, we must choose to harm the fewest. But, again, the reasoning is not utilitarian. To harm the many would grant greater weight to the moral rights of each individual in the small group, and lesser weight to the rights of each individual in the large group.

Now, suppose that the death of a mouse is comparable to the death of a human. (We will revisit this assumption below; we grant it here only for the sake of argument.) Suppose further that a million transgenic mice must be produced and killed in order to reap the benefit of the knowledge being sought in producing them. Would *MP* permit this experiment? Would it permit the production of the TAs previously mentioned?

It would not permit the production and slaughter of Beltsville hogs because such hogs are produced to gain knowledge that would lead to leaner meat carcasses. Providing humans with cheaper pork cutlets will not lead demonstrably to the saving of any human life. In general, therefore, *MP* will not justify the genetic engineering of experimental food animals because the harm to humans of foregoing the benefits of such research (cheaper meat) is trivial compared to the harm done to the animals. Since transgenic *food* animal research does not involve a trade-off of comparable harms, *MP* would not justify such research.

Neither would it justify the rearing and killing of animals such as the hair-loss mouse. The harm of slaughtering such a transgenic mouse is not comparable to the harm of failing to save a human the loss of hair. The loss of hair is a serious, even life-threatening, harm to a mouse. The loss of hair is a matter of mere vanity for many men. In general, therefore, the production of TAs for trivial purposes is not justifiable according to *MP,* because the harms in question are not comparable.

But, assuming again that the deaths of mice and humans are comparable, transgenic research that will demonstrably save the life of a human may be justifiable. Research on the shiverer mouse has identified the gene involved in causing the developmental defect of multiple sclerosis. Suppose that this research leads in a direct causal chain to saving the life of a patient with this disease. If the death of an animal is comparable to the death of a human, then *MP* commits AR theorists to the view that shiverer TA mice are justified when we know that producing and killing the mice will undoubtedly save the life of a victim of multiple sclerosis.

We have been operating under the assumption that the deaths of mice and humans are comparable. Regan does not grant this assumption, and neither should we. Defending the received intuition that the death of an ordinary human is always non-comparably worse than the death of an ordinary animal, Regan insists that *MP* does not apply to any of the cases under consideration. Another principle applies, the Worse-off Principle:

> *Where* **non-comparable** *harms are involved, avoid harming the worse-off individual (WP).*[69]

The justification of this principle follows from the principle of respect. Suppose I must choose one of only two options, either causing individual *N* to have a migraine or causing individual *M* to die. Since the harms involved are not comparable, the principle of treating *M* and *N* with equal respect entails that I choose to cause the lesser harm to *N*. Surely *N* would want me to make the same choice were the tables reversed, and I was contemplating having to cause *N* to die or *M* to have a headache.

Gary Varner illustrates the principle below, in Table 1.

Table 1: The Worse-off Principle, Case 1

Option # 1			**Option # 2**
-10	-10	-10	-1
-10	-10	-10	-1
-10	-10	-10	-1
-10	-10	-10	-1

Suppose we must either kill twelve humans or give four humans migraines. [70] Death is a catastrophic loss to each of the twelve, represented here as -10 (Option #1). A migraine is not a catastrophic harm, so it is represented here as -1 (Option #2). The Worse-off principle (*WP*) instructs us to choose Option #2.

In this case, the option recommended by a rights-theory dovetails with the option recommended by a utilitarian theory, because Option #2 has the consequence of minimizing overall harm. Whereas Option #1 entails a "harm score" of negative 120 (-10 x 12 = -120), Option #2 leads to a harm score of negative 4 (-1 x 4 = - 4). Option #2 is clearly the lesser of two evils.

But Regan insists that we should choose Option #2 not for the utilitarian reason but, rather, because it respects the rights of the worse-off. Choosing Option #1 would make any one of the twelve people in the left column worse-off than any one of the four people in the right column. Indeed, Regan argues, we must select Option #2, **even if** that choice will not have the consequence of minimizing overall harm. The reason is that we must avoid harming those individuals who will suffer the non-comparable, catastrophic, harms. Varner visually presents this consequence as follows:

Table 2: The Worse-off Principle, case 2

Option # 1	Option # 2				
- 10	- 1	- 1	- 1	- 1	- 1
	- 1	- 1	- 1	- 1	- 1
	- 1	- 1	- 1	- 1	- 1
	- 1	- 1	- 1	- 1	- 1
	- 1	- 1	- 1	- 1	- 1

If we must give migraine headaches to twenty-five people (-1 for each of the 25) in order to avoid killing one person (-10), we ought to do so, even though we will thereby cause a worse overall harm score of -25 compared to -10. In Table 2, Option #2 again prevails over Option #1, on the rights view, even though it entails causing more overall harm than Option #1. Protecting rights trumps the principle of minimizing overall harm.

To apply *WP* to animal research we must first review Regan's interpretation of harm. Harm is a diminution in one's capacity to form and satisfy desires. Because different individuals have different capacities to form and satisfy desires at different points in their lives, death can harm us to different degrees. Regan claims (p. 324) that the death of an ordinary human is never comparable to the death of an ordinary dog, but does not provide much argument for this claim in his eighth chapter. There are at least three reasons we may adduce in behalf of this claim.

First, the ordinary human exhibits a greater range, complexity, depth and sophistication of preference-interests than does the ordinary animal. We typically are capable of reflecting on our first-order desires and deciding to select some of them to pursue according to values and principles we have come to endorse. Consequently, we are morally free, and can try, for example, to adjudicate conflicting interests without the use of force. To the best of our knowledge, no other non-primate animals have such a capacity.

Second, our expected future life is typically longer and richer than the ordinary animal's expected future life. A young mother anticipates the adulthood of her child twenty or thirty years into the future; little evidence exists that any animals have this rich or extended a temporal horizon as they think about their offspring's future. Therefore, as Jeff McMahan has put it, death harms individuals in various ways, and:

> *The degree of harm an individual suffers by dying is a function of the net amount of good that the victim's life would contain if death were not to occur; and this in turn depends on the quality and quantity of the future goods the life would contain in the absence of death.*[71]

Notice that it is the quality of the goods lost, and not simply the quantity of them, that tells us how bad death is for someone. Killing a two-month old human fetus deprives it of more potential experiences than killing a twenty-three year old woman, but the death of the woman is a greater harm than the death of the fetus. The reason is that the woman possesses psychological unity, an individuality derived from her past choices and unique aspirations. The fetus lacks all psychological unity because it lacks a brain, brain waves, and experiences. As McMahan observes,

> *The extent to which an individual is harmed by the loss of some future good through death is a function both of the magnitude of the good and of the degree to which the person at the time of death would have been psychologically related to himself at the later time at which the good would have occurred within his life.*[72]

Third. The harm of death is not simply a function of the present or future satisfactions of an individual, nor is it a function of the total good an individual might realize later in life. It is also a function of the relation between the individual at the time they are killed, to the individual they would have been at a later point in time had they not been killed. The killing of a two year-old is worse than the killing of a two-month fetus because the two year-old has a unique psycho-social identity that will stand in a complex relation to the man the boy will be in twenty years. The fetus has no psycho-social identity and stands in nothing more than a physical relationship to the man the fetus will be in twenty years.

We may now apply this view of harm to the use of animals in research. As far as we know, mice lack not only moral autonomy, but the kind of memory and anticipation needed to give their lives long-term continuity and coherence. The relationship a mouse has to the individual it was a year ago is constricted because of its restricted memory, and the relationship the mouse has to the individual it will be a year from now is limited because of its dim sense of the future. Consequently, the harm that death would be to an ordinary mouse is not as bad as the harm that death would be to an ordinary young woman. The woman has a keen sense of where she has been and where she would like to go.

Discounting for these three factors need not lead us to conclude that death is not a harm to a mouse. Nor, certainly, that the random killing of mice is acceptable. It leads instead to Regan's conclusion: that the death of a human is noncomparably worse for a human than death is for a mouse and, when forced by circumstances to kill one or the other, we ought always to choose to kill the mouse.

WP will justify many cases of TA production if the experiment will save human lives and use animals that are not subjects of a life. For example, it is possible that a series of experiments involving the production and slaughter of transgenic mosquitos could provide the knowledge needed to achieve a total transmission blockade of the disease between insects. Were we able to introduce the antibody genes coding for anopheline into mosquitos, we could in theory express in the mosquito's midgut a protein activated by the mosquito's blood-sucking behavior that would block the transmission of malaria. Such a series of experiments might involve killing thousands of transgenic insects. Let us suppose that killing these TAs will directly save human lives. Here is an either/or choice: If we do not rear and harvest the transgenic mosquitos, we will have to stand idly by as humans die of malaria.[73]

Represent the harm of death to an ordinary adult human as a harm of the magnitude of -10.0. Represent the harm of death to an ordinary adult insect as, by comparison, less than -0.001, since insects presumably are not subjects of a life. In this case, the worse-off principle permits the sacrifice of thousands, or millions, of insects. Indeed, *WP* not only permits this option; it *requires* us to produce and kill vast numbers of such animals. Otherwise, we are not respecting the rights of the malaria victims.

Now, suppose that the experimental animal in question is not an invertebrate mosquito but a mammal with a comparatively highly developed brain and central nervous system. The harm of death to, say, a mouse will clearly be greater than the harm of death to a mosquito because of the greater range, complexity, depth and sophistication of the mouse's interests. If the harm of death to the insect is less than -0.001, and the harm of death to an adult human -10.0, then we might, somewhat arbitrarily, assign the harm of death to a mouse at -1.0. Because the harm of death to a human is still noncomparably worse than the harm of death to a mouse, *WP* will justify the same results for transgenic mice as for transgenic insects. *WP* instructs us that if the production of millions of transgenic mice will demonstrably save the live of one human being, then we ought to produce and kill those animals.

Admittedly, the consequence, insofar as it involves mammals, does not sound like one Tom Regan would endorse. Regan is known for his abolitionist stance toward the use of animals in science and, indeed, explicitly denies that his theory leads to the conclusion here described.[74] But why? Here we must distinguish two responses to this question: the response Regan gives in his book, and a response he provides later.

Consider his first response. Regan points out that *WP* begins with the clause "Special considerations aside . . . " He claims that a special consideration obtains in animal research rendering *WP* inapplicable to the case of animal experimentation. The special consideration is that the research animals are *innocent* individuals, and to kill them is to transfer risks to them against their consent. Regan believes we are never justified in transferring risks to innocent individuals against their will.

Regan's first attempt to block the application of the principle to animal research seems unconvincing for two reasons. First, the innocence of the animals is a difference that may make no difference. In *all* lifeboat cases, each individual is innocent by hypothesis. Were it not the case that all were innocent--were one of the individuals guilty of an offense that had caused the lifeboat dilemma to arise--then we would not have a lifeboat case. We would instead have a case with no dilemma, because the guilty party's offense would provide a good reason for preferring the death of the guilty party to the deaths of the innocent parties. Lifeboat cases are not like this. All parties are innocent; no individual has had any culpable causal role in the creation of the dilemma; and yet we still must choose to transfer the risk of death to someone. So, Regan's first attempt to block the applicability of *WP* to animal research seems to be a nonstarter.

Second, even if the innocence of the animals was relevant, the claim that we should never nonvoluntarily transfer risks to innocents (NTR) is not persuasive. As Gary Varner has pointed out in conversation, we often engage in NTR. Parents decide not to take a sick child to the clinic because they fear they cannot afford the health care.[75] The child recovers fully, and we do not blame the parents for their decision. Nonetheless, they have transferred risks to a child who was forced to accept them nonvoluntarily. Congress conscripts young men and women against their will into the army to defend the nation. The young people return to resume their careers, the war ends, and we do not blame Congress. A woman decides not to buy an insurance policy, thereby transferring to her dependents risks of loss of income that they probably would not have chosen to assume. All of these everyday occurrences are justifiable cases of NTR. So, Regan's second way of blocking the applicability of *WP* to animal research seems unpersuasive.

There is a third response open to Regan. In a 1985 exchange with Peter Singer, he wrote that "it is wrong--categorically wrong--coercively to put an animal at risk of harm, when the animal would not otherwise run this risk, so that others might benefit; . . . "[76] Here Regan appeals to the principle of respect, the foundation of the AR view, insisting that animal experimentation violates "the animal's right to be treated with respect by reducing the animal to the status of a mere resource, a mere means, a thing." Because mammals used in research are subjects of a life, seizing them against their will and conscripting them into painful or lethal research fails to show due consideration for their negative right not to have the integrity of their bodies and projects violated. Thus, *WP* is inapplicable to the case of using animals in research because one of its conditions is not satisfied: there is, at the beginning, no conflict of rights involved. An apparent conflict of rights is *generated* by an agent's overriding an innocent animal's rights.

This response is powerful, and raises an issue of central importance. The issue is that there is a fundamental disanalogy between the case of animal research and lifeboat cases. As lifeboat cases are much discussed in the literature of applied ethics, this disanalogy deserves careful exploration.

Standard lifeboat cases have the following requirements:

(1) An agent, *A*, whose own rights are in danger of being violated, is required to choose between violating her own rights or the rights of one or more others (call these others, "stakeholders");

(2) The other stakeholders (*B*, *C*, and so on) are determinate;

(3) None of the stakeholders has been placed in harm's way by the actions of *A*.

(1) requires that the agent facing the lifeboat dilemma must herself be in danger of severe harm if no choice is made. (2) requires that the stakeholders be definite, neither unknown victims still to be determined or faceless representatives of future generations. The stakeholders are persons *A*, *B*, and *C*, that is, specific individuals in the lifeboat with *A*. (3) requires that the

presence of each stakeholder in the lifeboat is free, that is, none were coerced by *A* to be on the lifeboat.[77]

None of these requirements is satisfied in the case of animal research. The researchers who decide to use animals in research typically are not themselves victims of the disease they are hoping to cure. The animals are not determinate individuals; we cannot name the dogs ahead of time that will be selected for use. And the dogs do not come naturally, as it were, into the experiment; they are forced into it.

For these reasons, we ought to abandon the lifeboat metaphor in trying to think through the case of animal experimentation. A more apt analogy is a version of a trolley-car case.

The standard trolley-car case describes an individual faced with a moral dilemma in which the individual herself will not be faced with serious harm if no choice is made. The individual nonetheless must choose between an action that will inevitably lead to the killing of a group of innocent individuals, or taking no action at all, which will result in the killing of a different group of innocent individuals.

Imagine a trolley-car careening down a hill out of control. It approaches a switch between track #20 and track #3, a switch over which you have control. On track #20 are twenty innocent humans. On track #3 are three innocent humans. If you do nothing, the car will continue on track #20, killing twenty humans. The three humans on track #3, of course, will remain unharmed. If you flip the switch, however, redirecting the car onto track #3, you will directly cause the death of three people, while directly saving the lives of twenty people.

Now, imagine the following permutation on the case. As you look down track #3, you do not see three people on the track. You see a large carousel carrying hundreds of vacationers. When at rest, the carousel does not intersect the path of track #3 and no one on the carousel is at risk. Once started, however, the gyrating motion of the carousel intermittently carries part of it directly over track #3. At any given moment, the circling carousel rotates three passengers into the path of track #3. The three people in the fateful location are "indeterminate," in the sense that a different set of people is at risk at each moment. We cannot identify ahead of time just who exactly will be put at risk. The constitution of the risk pool, in other words, changes from second to second.

Imagine, finally, that a faulty wiring job has hooked the carousel's start switch to the trolley-car switch in your hands. You must, therefore, make the following choice. Do nothing, and allow the trolley-car to kill twenty innocent people on track #20 while not endangering anyone on the carousel. Or, flip the switch, redirecting the trolley-car toward the carousel, and saving the lives of twenty specific people on track #20 while simultaneously setting in motion a process that will kill three, as-yet-undetermined, people-about-to-be-rotated onto track #3.

This case has the following features:

(4) An agent, *A*, whose own rights are not in danger of being violated, is required to choose between doing nothing, with the result that the rights of many determinate individuals are violated, or doing something, with the result that the rights of a few, other, individuals are violated (call these individuals the "stakeholders"),

(5) The identities of the stakeholders in the smaller risk pool are not determinate;

(6) The stakeholders in the smaller risk pool are directly put into the risk pool by the actions of *A*.

Our intuitions seem clear, and reliable, here; we ought to flip the switch and kill three rather than twenty. This intuition is justified by Regan's theory, because we have here a case of deciding between comparable harms, as the death of one human is comparable to the death of another. Therefore, the minimize overriding of rights principle applies, and we ought to choose to violate the smallest number of individuals' rights as possible.

Is the trolley-car case also analogous to (some) cases of animal research? Imagine that the individuals on Track #20 are humans, and the individuals on the carousel are animals. The harm of death to a human is non-comparably worse than the harm of death to an animal, so the mini-ride principle will not apply here. But *WP* will apply, and will justify choosing to kill the three animals, if the features of the case really are analogous. Are they?

In animal research, the researcher's own rights typically are not in danger of being violated because researchers usually do not suffer from the disease they are trying to cure. Yet the researcher is required to choose between doing nothing, with the result that the rights of many diseased innocent humans will be violated, or doing something, namely, killing animals. The result of the second choice will be that the rights of other innocent individuals, the animals, will be violated. Condition (4) is satisfied.

The stakeholders in animal research are not determinate. This is obvious in the case of pound seizure. When selecting dogs for research from stray and abandoned animals, we clearly do not know ahead of time which pool of animals will be selected, because we do not know which animals will come to be in the pound at any moment. We may not even know which pound will be selected. Therefore, we may think of all of the stray and abandoned dogs in the US as riders on the carousel. Exactly three of the dogs will be selected and harmed, but we have no way to identify at present *which* three they will be. Condition (5) is satisfied in the instance of random selection of animals for experimentation.

Pound seizure is no longer the technique of choice, however, and the vast majority of animals used in research are intentionally bred for this purpose. Is (5) satisfied when researchers directly set out to bring lab mice into existence?

I think so. Imagine that each mouse has a soul and before it is born, God drops its soul into this or that mouse embryo. As a result of this natural lottery, some mice are born into laboratories, others into haymows. Those born into labs are unfortunate, and will come inevitably into harms' way at the hands of humans. The others will not; they will have to fend only with barn cats. If this thought-experiment is defensible, then condition (5) is satisfied even in the case of intentional breeding of lab animals.

Finally, while animal research does not require that researchers initially determine the identities of the animals they are going to use (God does that), animal research clearly does require researchers (or their delegates) directly to select animals to be put in harm's way. Someone must enter the pound, survey the room of candidates, and then select one animal for experimentation while sparing others. Someone must directly oversee the selection of individual mice whose sperm and eggs will be combined to produce the embryos to be brought to term and then used for experimentation. Some stakeholders are put into the risk pool by our actions. Condition (6) is also satisfied.

Unlike abolitionists who oppose all TA production, those who adopt ARR must endorse the production of some TAs in order to abide by the provisions of *WP*. Under certain conditions, research may have the direct effect of saving the lives of human beings who would be made worse-off in the absence of the research. It bears noting that this condition probably applies to a small minority of the actual cases of the use of TAs in scientific research. Be that as it may, the carousel case suggests that the production of TA neoplasmic mice, and even dogs, may be justified according to the worse-off principle. If we must make a choice, either to do nothing, and so watch as a group of humans dies, or produce and kill indeterminate research animals with the result that the humans are saved, then we ought to violate the rights of the animals. This is true even if:

- all of the animals are innocent;
- we do not know exactly which animals will be harmed; and,
- we must directly select the animals to be put into the risk pool.

I conclude that an animal rights ethic is not committed to an abolitionist approach to the use of animals in research. Let us consider another possibility.

> ARR: *An animal rights ethic must oppose the use of animals in research and agriculture whenever the harm to be done to the animal is noncomparably greater than the harm that would be done to the human were the animal not subjected to harm.*

According to ARR, then, the production of hairless mice is not justified, since the harm to the animal is noncomparably greater than the harm that would be done to the balding man were the hairless mouse not produced. Nor is the production of transgenic pigs to be used for slaughter justified, since the harm that death is to the pig is noncomparably greater than the harm that would be done to the consumer were the consumer denied a veal cutlet.

What follows from ARR for the propriety of producing transgenic farm animals (TFAs)? Five points should be made.

First point. *ARR does not protect animals that lack preference-interests.* There are, of course, other moral concerns to be considered, including the potential ecological consequences of TA production and questions about the influence of the work on the character of those doing the scientific research. However, in the absence of other moral reasons not to genetically modify them, the following phyla are *prima facie* eligible for experimental manipulation: Annelida (earthworms and leeches), Echinodermata (starfish, sand dollars, sea urchins, sea cucumbers), Mollusca (clams, shrimp, oysters, scallops), and Arthropoda (insects, crustaceans, spiders, centipedes).[78] If a transgenic animal is incapable of having a future, then it is *impossible* to do direct harm to it by killing it painlessly.

Where do we draw the line? Rights are tied to interests, so animals with an interest in not feeling pain have, all other things being equal, a right not to be caused pain for a trivial reason. Animals with a future, have, all other things being equal, a right not to be killed for a trivial reason. It is one thing to try to determine whether an animal is sentient. It is another thing to try to determine whether an animal has a future. How do we do that?

Following pioneering work of Gary Varner, I want to suggest that the answer is to look for signs of hypothesis formation and testing. The reason is that "having a future" consists in more than mere reflexive responses to avoid adverse stimuli, a "behavior" of which many plants are capable. Having a future means having a desire; being capable of formulating at least one hypothesis as a way of satisfying that desire; and subsequently choosing to act on one's hypothesis.[79] By formulating and acting on hypotheses, individuals can learn; that is, they can assimilate and store knowledge about past successes and failures so as to improve their ability to form future hypotheses. Having a future means, in sum, being able to shape one's future.

Which animals have the capacity to shape their future? At some point in pre-history, the universe witnessed the first animals capable of proto-reasoning about their desires, and these animals were the first individuals with futures. They must have had a sufficiently developed central nervous system and brain to support the complex mental operations required for hypothesis formation and testing.[80]

By hypothesis formation and testing I, like Varner, do not have in mind anything very intellectual. I take it that an individual's behavior can be explained in terms of hypothesis formation and testing even when the individual is incapable of articulating that they are engaged in forming and testing hypotheses. A coyote has to decide to act on one of two competing desires: (a) to attempt to feed its offspring or (b) to sleep. The coyote may hypothesize that by searching first for food she will be able to sleep later as well, but that if she sleeps first, prey will escape, her offspring will starve, and she will have failed to feed them. Her hypothesis, though not of course at this level of abstraction, is that by acting on (a) she will also have a good chance to act on (b), but by acting initially on (b) she will forego the opportunity to act on (a). So she tests the hypothesis by acting on (a) rather than (b). The coyote is involved, at least, in proto-reasoning about her desires.

The empirical task, then, is to examine the behavioral and the physiological evidence for different species, and to try to determine which species are capable not simply of movement, which can be attributed to instinct and habit, but of the kind of learning involved in hypothesis formation and testing. Varner's review of the available empirical evidence leads him to this conclusion:

> Fish and lower animals almost certainly do not have desires. Mammals almost certainly have desires, and in them, the practical reasoning characteristic of desire is localized in the prefrontal cortex. Birds probably have desires (although the case for saying that they do is somewhat weaker than that for saying that mammals do), and in birds, practical reasoning is localized in the hyperstriatum. Reptiles may have desires (although the case for saying that they do is decisively weaker than that for saying that birds do), and if reptiles have desires, the related practical reasoning is localized somewhere in the primitive reptilian cerebrum.[81]

How does this line of reasoning help us with the line-drawing problem? The moral obligations scientists have to TAs will depend on the level of sentience and consciousness possessed by the animals with which they are working. It is wrong to cause pain for trivial reasons to any sentient animal, including fish. The reason is simply that pain is bad and 10 units of pain are 10 units of pain whether I suffer them, or you suffer them, or a rainbow trout suffers them. Here it does not matter that you and I can desire for the pain to cease in the future, whereas the trout apparently cannot. Thus, all TA research on sentient animals should be bound by the Worse-off rule.

But research on *animals that lack futures* need not be bound by the "No harvest TA" rule that I proposed at the end of chapter 3, because these animals do not possess the characteristics needed to be entitled to the right not to be killed painlessly.

Second point. *ARR recognizes the right to life of all animals who have a future.* As noted in ch. 3, the transgenic animal of choice is the mouse, a vertebrate with a complex brain and nervous system. The slaughter of a research mouse interferes with its capacity to pursue its interests and shape its future. Scientists working with this and other vertebrate species need not necessarily call a moratorium on all of their research, but the burden of proof is on them to show that their research will directly save human lives. Any lesser goal is probably morally unjustifiable.

Third point. *ARR does not justify the production and slaughter of animals for food in developed countries.* It is rarely the case in developed countries that one must choose between the life of a so-called food animal and the life of a human. As Peter Singer puts it, eating meat is a "great extravagance."

> Some 38 per cent of the world's grain crop is now fed to animals, as well as large quantities of soybeans. There are three times as many domestic animals on this planet as there are human beings. The combined weight of the world's 1.28 billion cattle alone exceeds that of the human population.[82]

Like ARA, the more permissive ARR theory of animal rights will not allow suffering or slaughter either of research or production TFAs unless the research can be directly linked to the saving of lives, perhaps in developing countries. But transgenic farm animals would not seem to be the answer to the problem of starvation.

Fourth point. *To the extent that TFAs are not slaughtered or harmed, TFA research may be justified.* It seems that much progress in controlling the regulator genes has been made since the 1985 Beltsville hog experiments with the result that few transgenic research animals seem to suffer as a result of physiological problems.[83] Indeed, if the reports of those producing TFAs in the public sector are to be believed, TFA suffering seems to have been virtually eliminated.[84]

One reason is that embryos with significant defects are not brought to term. In producing the first cloned sheep, Ian Wilmut destroyed nearly 300 embryos before successfully producing Dolly. Unless one holds that embryos are capable of suffering, this research, while "wasteful" in one sense, did not cause the kind of suffering seen in the Beltsville hogs case. Another reason is that inserted genes are now targeted to work in specific organs, such as the mammary gland, rather than in central "unregulated" physiological systems, such as the growth hormone system. As a result, the vast majority of TFAs produced in the last ten years have grown normally with minimal if any side effects. One example: dairy researchers today are not inserting genes for greater milk production into all of the cow's cells; they are rather targeting genes to work in more local, specific, ways by introducing them directly into the cow's mammary gland. Therefore, only the cells in one of the cow's organs have the foreign gene in them.

While there is no hard data on the extent of animal suffering in research labs producing TFAs, the judgments of those working in this area appear to be fairly uniform. As the previous footnote attests, the majority of researchers seem to believe that there is little suffering among current TFAs.

Fifth point. *The right target at which to aim when concerned to protect sentient animals may be the legal system rather than the research system.* Protocols have been developed requiring researchers to hold animal pain to a minimum. The Beltsville hogs were subjected to unnecessary suffering insofar as those running the experiment presumably could have given the arthritic hogs an analgesic once the animal's condition was discovered. Or, the scientists could have euthanized the piglets after diagnosing their problems. The scientists' failure to do so was legally justified by the fact that farm animals are exempt from the provisions of the Animal Welfare Act, and scientifically justified, one presumes, on the basis that analgesia might have compromised the reliability of the experimental results. Nonetheless, such gaps in the regulations can and should be fixed.

Global critics of ag biotech shoot at the wrong target in trying to stop all animal biotechnology. It is possible to eliminate most if not all of the suffering of transgenic animals by legislating against procedures likely to produce it. European laws, for example, require researchers to weigh the amount of animal suffering in an experiment against the expected benefits to humans.[85] While there is ample room in the United States for legislation to prohibit experimentation that might cause suffering in animals capable of suffering, there is still much to be done in animal law. The US has large loopholes in its legislation, and no single unitary scheme covering all animals.

In sum, the animal rights view I now hold has very different attitudes to the use of animals in agriculture and their use in research. ARR takes an abolitionist approach to agriculture, and a reformist attitude to animal experimentation.

No slaughter of animals with futures for meat in developed countries.

Carefully circumscribed use of animals when research can save determinate human lives.

Conclusion

We began by considering the following proposition.

> *If there is a substantial risk that a technology will do more harm than good to humans, ecosystems, and animals, then it should not be developed.*

We assumed that this proposition is true, then noted that the strength of the extrinsic case against ag biotech hinges on the idea that potential harms to humans, land, and animals will be greater than the benefits.

We examined eight arguments for this conclusion and found several areas in which we must continue to be concerned about the release of GMOs. These areas include: The EPA exemption for viral coat proteins and the risks that this policy entails for the appearance of new viral plant diseases **(1)**. The difficulties facing developing countries as they try to negotiate the legal and financial mazes of intellectual property rights **(3.3)**. The environmental risks posed by gene-flow in crops grown near weedy relatives **(6.6)**. The possibility that transgenic fish might cross-breed with, or outcompete, wild salmon, with the resultant loss of wild species **(6.7)**. The use of animals with futures in research aimed at trivial goals, such as restoring men's hairlines **(8.2)**.

These and other risks are not to be taken lightly. It appears, however, that we may be able to minimize them through regulatory action. Overall, when weighed against the potential benefits of the new genetics, the risks outlined in this chapter do not add up to a vindication of the global case against ag biotech.

Notes

[1] Cf. Kristin Shrader-Frechette, " Reductionist Approaches to Risk," in Deborah G. Mayo and Rachelle Hollander, *Acceptable Evidence: Science and Values in Risk Management* (New York: Oxford University Press, 1991), pp. 218-248.

[2] Zeckhauser and Viscusi, *Science* 248, pp. 559-564. Cited in Jan Leemans, "Safety in Biotechnology: Putting the Data in Context," at: http://nbiap.biochem.vt.edu/articles/MAR9110.htm.

[3] Monsanto tested genetically engineered tomato plants resistant to virus, insects, and Roundup herbicide on a farm near Jerseyville, Illinois, in June 1987. According to the Monsanto *Backgrounder* (May 1989), this was the "first genetically engineered food crop allowed by regulators to flower and bear fruit."

[4] Nordlee, J.A., Taylor, S.L., Townsend, J.A., Thomas, L.A. and Bush, R.K. "Identification of brazil-nut allergen in transgenic soybeans," *The New England Journal of Medicine* (March 14, 1996): 688-728.

[5] Slutsker L et al, "Eosinophilia-myalgia syndrome associated with single tryptophan manufacturer," *JAMA* Jul 11; 264:213-217. Abstract published in *Journal Watch* 17 July 1990.

[6] John Fagan, *Genetic Engineering: The Hazards, Vedic Engineering: The Solutions* (Fairfield, Iowa: Mahirishi International University Press, 1995), p. 54. Fagan cites A. N. Mayeno, and G.J. Gleich, "Eosinophilia-myalgia syndrome and tryptophan production: a cautionary tale," *Tibtech* 12 346-352, 1994.

[7] Belongia, E.A., Hedeberg, C.W., Gleich, G.J., White, K.E., Mayeno, A.N., Loegering, D.A., Dunette, S.L., Pirie, P.L., MacDonals, K.L. and Osterholm, M.T. "An investigation of the cause of eosinophilia-myalgia syndrome associated with tryptophan use," *The New England Journal of Medicine* (1990) 323 (6), 357-364. Cited in Jens Katzek, "Product Contamination: The L-Tryptophan Case," Center for Global Food Issues, at: www.cgfi.com/myths_det.cfm?MID=6

[8] Raphals, P., "Does medical mystery threaten biotech?" *Science* 249 (1990), 619. Quoted in Fagan, "The Facts about Genetic Tryptophan: a Summary," November 1997. On the web at http://www.zmag.org/Bulletins/ptry.htm

[9] Gary P. Munkvold, "Comparison of fumonisin concentrations in kernels of transgenic Bt maize hybrids and nontransgenic hybrids," *Plant Disease* 83 (1999): 130-138.

[10] Many thanks to Steven R. Shafer for explaining the risks and sources of naturally occurring insecticides to me and for doing yeoman's work in reviewing this section of the chapter. With his permission, I have borrowed freely from his presentation, coauthored with Michael D. McElvaine, and Alwynelle S. Ahl, "Ethical Issues in Risk Communication: Why Consumers Need Not Worry About Genetically Modified Crops," unpublished paper presented at Iowa State University Winter Bioethics Faculty
Retreat, Ames, Iowa, January 2000. Available at www.biotech.iastate.edu/Bioethics/gmosethics/USDA.pdf.

[11] See "Table 1: Glycoalkaloids in common foods for 20-kg child (compiled by L. C. Abbott and S. R. Shafer, USDA Office of Risk Assessment and Cost-Benefit Analysis, Washington, DC), at: www.biotech.iastate.edu/Bioethics/gmosethics/USDA.pdf.

[12] See www.aphis.usda.gov/biotechnology/index.html.

[13] Federal Register 51:23302-23350.
[14] "Genetically Modified Pest-Protected Plants: Science and Regulation," available from the National Academy Press, 2000. "U.S. Regulatory System Needs Adjustment As Volume and Mix of Transgenic Plants Increase in Marketplace," is available on line at: www.nas.edu/.
[15] James E. Schoelz and William M. Wintermantel, "Expansion of Viral Host Range through Compelmentation and Recombination in Transgenic Plants," *The Plant Cell* 5 (Nov 1993): 1669-1679. A. E. Greene and R. F. Allison, "Recombination between Viral RNA and Transgenic Plant Transcripts," *Science* 263 (1994): 1423-1425. Thanks to Bryony Bonning for bringing these references to my attention.
[16] Lynn R. Goldman, "Introductory Lecture on Biotechnology Issues," lecture presented at the Fourth Annual Shapiro Environmental Law Conference, "Biotechnology and the Human Environment," The George Washington University Law School, 13 April 2000.
[17] The National Academy's report confirms the complaints of ag biotech's opponents, that the agencies are not transparent enough.
[18] Vernon W. Ruttan, "Biotechnology and Agriculture: A Skeptical Perspective,"*AgBioForum* 2 (Winter 1999), pp. 1-7, citation at p. 3. On web at: www.agbioforum.missouri.edu/AgBioForum/vol2no1/ruttan.html.
[19] Thanks to my colleague Clark Ford for information found in the next two paragraphs. See Ford, "Ethics and GMOs," powerpoint presentation, at: www.biotech.iastate.edu/Bioethics/gmosethics/ford.pdf.
[20] Guy Gugliotta, "New Rice Strain Termed Nutrition Breakthrough," *Washington Post*, Friday, January 14, 2000; Page A6. On web at: http://washingtonpost.com/wp-dyn/health/A43053-2000Jan13.html
[21] Potrykus presented his results at a conference called AGBIOTECH 99: Biotechnology and World Agriculture in London, November 1999. His research results were published in *Science* 287 (2000): 303-305, 241-243. Cf. www.agbiotechnet.com/topics/devco.asp.
22. At least twenty potential vaccines for livestock animals have been produced using genetic engineering. *Chemical & Engineering News*, April 14, 1986, p. 4; April 21, 1986, p. 7; April 28, 1986; *New York Times*, April 13, 1986. Articles cited by Weppelman, in Vasil, p. 67.
[23] Carl Feldbaum, "My Brush with Biotech--As a Fortunate Patient," *BIO News* (Feb / Mar 2000): 1, 3.
[24] Being developed by Schering-Plough, Abbott Laboratories and SmithKline Beecham. George S. Mack, "Biogen converts healthy sales into surging stock," at moneycentral.msn.com/articles/invest/company/3134.asp
[25] U. S. Bureau of the Census, *1992 Concentration Ratios in Manufacturing* (1996); as quoted in Fred M. Gottheil, *Principles of Economics, 2nd ed.* (: , 1999), p. 253. "In 199 of the 448 industries (that's 44.4 percent of the industries, the four leading firms control at least 40 percent of sales." Thanks to Rich Noland for calling my attention to the prevalence of oligopoly.
[26] Michael N. Kibue, "A Farmer's View," Joske Bunders, Bertus Haverkort and Wim Kiemstra, editors, *Biotechnology: Building on Farmers' Knowledge* (London: Macmillan Education Ltd., 1996), pp. 15-22, citation at p. 19.
[27] Kibue, p. 20.
[28] Partha Dasgupta, "Population, Poverty, and the Local Environment," *The Ag Bioethics Forum* 7 (1995): 1-5, citation at p. 3. The article is reprinted from *Scientific American* 272 (1995): 26-31. Also see Dasgupta, *An Inquiry into Well-being and Destitution* (New York: Oxford University Press, 1993).

29. Crosson, Pierre Crosson, "Sustainable Food Production: Interactions among Natural Resources, Technology, and Institutions," *Food Policy* (May 1986): 153. Crosson refers to World Bank, *Commodity Trade and Price Trends* (Baltimore: Johns Hopkins University Press for the World Bank, 1981), Ref 15, p. 70.
30. Crosson, "Sustainable," 153. Crosson cites International Rice Research Institute, *Changes in Rice Farming in Selected Areas of Asia* (Laguna, Philippines: Los Banos, 1975); and Y. Hayami and M. Kikuchi, *Asian Village Economy at the Crossroads* (Baltimore: The Johns Hopkins University Press, 1975).
31. Crosson, "Sustainable," 154.
32 R. S. Swaminathan, address to Regional Seminar on Public Policy Implications of Biotechnology for Asian Agriculture (New Delhi, India, 1989), quoted in Duvick, "Is Biotechnology Compatible with Sustainable Agriculture? Yes," in *The Ag Bioethics Forum* 4 (June 1992), p. 4.
33 Stephen Goldman, who is working to breed asexually reproducing corn, has been quoted as saying that this "is the challenger to the terminator technology." Andrew Pollack, "Looking for Crops that Clone Themselves," *The New York Times,* Tuesday, April 25, 2000, p. D3.
34. Referee's report on "The Case Against bGH," Society for Philosophy and Technology, March 1990. On the problems of large producers competing unfairly with smaller producers, and of squelching innovations, the reviewer directed attention to Walter Adams and James Brock, *The Bigness Complex*; David Noble, *America by Design*. On the relationship between technology and participatory democracy, the reviewer referred to Carol Goul, *Rethinking Democracy*; Joshua Cohen and Joel Rogers, *On Democracy*; and, for works in political economy concerned with democratic theory, Samuel Bowles and Herbert Gintis, *Capitalism and Democracy*; Bowles, Gordon and Weisskopf, *Beyond the Wasteland*; and Martin Carnoy and Derek Shearer, *Economic Democracy*.
35 Bryan, Lowell, and Diana Farrel, D, *Market Unbound: Unleashing Global Capitalism* (New York: Wiley & Sons, 1996). Thanks to my colleague Tony Smith for directing my attention to this book, and for assistance in thinking through some of the problems with globalization and biotechnology. I highly recommend his essay, "Biotechnology and Global Justice," forthcoming in *Journal of Agricultural and Environmental Ethics.*
36 Korten, David, *When Corporations Ruled the World.* (New York: Kumarian Press, 1996).
37 C.S. Prakash, forthcoming in *Technology Review* (MIT Press). Prakash believes that we need "an independent middleman . . . a new international foundation [to] . . . assume responsibility for transferring technology to developing countries . . . [by] indemnify[ing] companies from liability suits that may arise from the use of their donated technologies, [and helping to] negotiate the labyrinth of patent laws, intellectual property claims and country-specific concerns."
38 Thanks to Humberto Rosa for suggesting this intriguing possibility.
39. Crosson, "Sustainable," 143.
40. Crosson, "Sustainable," 144.
41. Crosson, "Sustainable," 145.
42. Crosson, "Sustainable," 144.
43. Crosson, "Sustainable," 144.
44. Wendell Berry, "The Making of a Marginal Farm," in *Recollected Essays: 1965-1980* (San Francisco: North Point Press, 1981): 340.
45. Crosson, "Sustainable," 144.
46. Crosson, "Sustainable," 145, citing a study by S. Aldrich, *Nitrogen in Relation to Food, Environment and Energy*, University of Illinois, Champaign-Urbana, IL, 1980.

47. D. Pimentel, et. al., "Environmental and Social Costs of Pesticides: A Preliminary Assessment," *OIKOS* 34 (1980): 126-140.
48 Don Duvick, "Is Biotechnology Compatible with Sustainable Agriculture?" *The Ag Bioethics Forum* 4 (June 1992): 3.
49 P. T., "Release of Recombinant Microbes: Data from the Field," (January 1996), reporting on a study at the University of Surrey using a strain of Pseudomonas fluorescens. At: http://nbiap.biochem.vt.edu/articles/jan9610.htm.
50. Brill, p. 80.
51 www.monsanto.com/ag/articles/PlantBiotech/
52 Cf. Ellen Messer, "Agricultural Biotechnology: Potential Promise and Peril," at http://www.brown.edu/Administration/George_Street_Journal/v20/v20n23/messer.html
53 Thomas E. Nickson and Michael J. McKee, "Ecological Aspects of Genetically Modified Crops," in Ralph W. F. Hardy and Jane Baker Segelken, *NABC Report 10, Agricultural Biotechnology and Environmental Quality: Gene Escape and Pest Resistance* (Ithaca, New York: National Agricultural Biotechnology Council, 1998), p. 103.
54 Jane Rissler and Margaret Mellon, *Perils Amidst the Promise: Ecological Risks of Transgenic Crops in a Global Market* (Washington, D.C.: Union of Concerned Scientists, 1993). On-line at: www.cbi.pku.edu.cn/binas/Library/ucs/
55 Rissler and Mellon, op cit. On-line at: www.cbi.pku.edu.cn/binas/Library/ucs/section5.3.html.
56 Kirsty Mclean, "Commercial Release of GMOs in Australia," February, 1996, at http://nbiap.biochem.vt.edu/articles/feb9605.htm.
57 Andrea Johanson, *Assistant Director*, ABSP, MSU, www.iia.msu.edu/absp/biosafety.html
58 Johanson.
59 Federation of American Societies for Experimental Biology, Life Sciences Research Office, "Evaluation of the Agricultural Biotechnology Risk Assessment Research Grants Program: Comments from the Scientific Community and Recommendations for Future Programming," p. 6. At http://nbiap.biochem.vt.edu/faseb/faseb.toc.html.
60 For a more pessimistic assessment, see Margaret Mellon, "ABRAC Guidelines Show the Inability of the Government to Regulate Deliberate Release," http://nbiap.biochem.vt.edu/articles/jun9104.htm.
61 Email message from Etis dated 4/24/00. Web site at: http://afprotein.com.
62. National Resesarch Council, Board on Agriculture, Alternative Agriculture (Washington, DC: National Academy Press), quoted in Sandra S. Batie, "Agricultural Policy and Environmental Goals: Conflict or Compatibility?" *Journal of Economic Issues* 24 (June 1990): 565-573, p. 567.
63. Polly Ligon, Sandra S. Batie, Waldon R. Kerns, Daniel B. Taylor, and Paul B. Siegel, "Chesapeake Bay Farmers Participation in the Conservation Reserve Program," unpublished document, Department of Agricultural Economics, Virginia Polytechnic Institute and State University, Blacksburg, Virginia, 1988; cited in Batie, "Agricultural Policy," p. 569.
64 G. F. Warren, "Spectacular increases in crop yields in the United States in the twentieth century," *Weed Technology* 12 (1998): 752-760. Thanks to Robert Zimdahl for calling my attention to this reference in his manuscript, "Moral Certainty in Agriculture."
65 Dennis Avery, "Saving the Planet with Pesticides and Biotechnology and European Farm Reform," Pages 3-18 in British Crop Protection Conference - Weeds (1997). Quoted in Zimdahl, "Moral Certainty in Agriculture."
66 "Panda Cloning Makes Breakthrough," *Associated Press, Copyright 1999.* On web at: http://www.discovery.com/news/briefs/brief3.html

67 Regan, *Case*, p. 325.
68 Regan, *Case*, p. 305.
69 Regan, *Case*, p. 305.
70 I have drawn extensively in this section from ideas presented in Varner's important book, *In Nature's Interests? Interests, Animal Rights and Environmental Ethics* (Oxford University Press, 1998) and on his web page, "A Lecture on Animal Rights and Animal Welfare." The charts reproduced here are at http://snaefell.tamu.edu/~gary/awvar/regan_2.html
71 Jeff McMahan, "The Moral Status of Animals," unpublished manuscript, presented at the Iowa State University Model Bioethics Institute, the University of Illinois, May 1994.
72 McMahan, "Animals," p. 1.
73 Transgenes having been injected into the fruitfly *Drosophila melanogaster* since the 1960s. No record is kept of the number of transgenic insects produced, but given the global nature of the insect research establishment, the speed of insect reproduction, and the size of new insect generations, it is probably true that billions of transgenic insects are produced every year. Julian M. Crampton and Paul Eggleston, "Transgenic insects," in Norman Maclean, *Animals with Novel Genes* (Cambridge: Cambridge University Press, 1994), p. 21-23.
74 Regan, *Case*, p. 377.
75 Private conversation.
76 Peter Singer and Tom Regan, "Dog in the Lifeboat: An Exchange," *New York Review of Books* (April 25, 1985), p. 57.
77 For an important discussion of these issues, see Dale Jamieson, "Rights, Justice, and Duties to Provide Assistance: A Critique of Regan's Theory of Rights," *Ethics* 100 (January 1990): 349-362.
78 Regan thinks AR protects at least all normal adult mammals of one year or older and, probably, most other vertebrates (Case, pp. 78 and 367). Gary Varner has offered an intriguing and detailed defense of the view that while mammals and birds probably have desires, fish and invertebrates probably do not. In chapter two, "Localizing Desire," of *Vexing Nature*, Varner summarizes findings about the distribution of basic learning capacities among animals. As Varner argues on his web site, "these capacities seem relevant to possession of desires insofar as it seems implausible to attribute conscious planning to organisms which lack the capacities in question." His conclusions are based in turn on Martin Bitterman, "The Evolution of Intelligence," *Scientific American* 212 (1965), pp. 92-100, and Varner's review of subsequent research.
79 I am indebted to Varner for the idea that the best way to understand what is wrong with killing animals is by paying attention to their abilities to learn, to form and test hypotheses. See Varner, *In Nature's Interest?* pp. 30-37.
80 To develop such brain structures and neural pathways in the hominid, similar structures must have evolved in the hominid's precursors. As Holmes Rolston suggests, the capacity for choice must have "evolved out of choicelike precedents in the protopsychologies of animal behavior. "Cf. Holmes Rolston, III, *Environmental Ethics: Duties to and Values in The Natural World* (Philadelphia: Temple University Press, 1988), p. 23.
81. Varner, p. 76. Varner refers in an endnote to the work of M. E. Bitterman, "The Evolution of Intelligence," *Scientific American* 212 (January, 1965), pp. 92-100, and comments that "Bitterman did not do lesion studies to determine where in the reptilian cerebrum the ability for progressive adjustment is localized. But given that in both mammals and birds, the ability is localized somewhere in the cerebrum (rather than in, say the midbrain), it is reasonable to assume that the ability for progressive adjustment is localized somewhere in the cerebrum of the reptiles, from whom both the birds and the mammals evolved" (p. 84, note 29).

82 Peter Singer, *Practical Ethics, 2nd ed.* (Cambridge: Cambridge University Press, 1993), p. 287.
83 As of Fall, 1997, there were five groups actively producing TFAs: Genzyme Transgenics, Inc. (goats, for milk with protein pharmaceuticals; PPL in Blacksburg, Virginia (cows, for milk with a protein to be used in baby food formulae; USDA Beltsville ARS (pigs, for leaner meat) and Advanced Cell Technology, Inc. in Massachussets.
84 In preparing this manuscript, I asked a dozen prominent TFA researchers two questions: (a) What is the total number of TFAs produced annually? And (b) What percentage of TFAs produced annually are significantly worse-off than they would have been had they not been tampered with at the embryonic stage?

Results of my unscientific survey showed that there is no central clearinghouse for the information sought in (a). Therefore, calculating the percentage sought in (b) is impossible. But of the researchers who hazarded guesses, all agreed that the number of transgenic cattle could be probably counted on one's fingers; that the number of founder pigs would be be less than 200, and that goats and sheep would be in the thousands. (Thanks to my colleague, Gary Lindberg, for these guesses). While I did not provide them with a clear definition of what "being significantly worse-off" means, none of them seemed to think that the concept was fuzzy or difficult to understand. And they all agreed that the percentage in (b) is probably very low. Indeed, while all of the reporters were cautious, most seemed to think that the Beltsville hogs were the only transgenic animals that suffered as a result of genetic engineering. They gave four reasons.

a) *Embryos don't suffer*. The result of physical damage to the embryo is almost always pre-natal lethality, so few transgenic large animal embryos are actually brought to term. Results of transgene effect may be construct dependent or integration site dependent. Lines with deleterious phenotypes as a result of transgene integration site are almost always discarded, unless the phenotype is of scientific interest in its own right. This has happened occasionally in mice but apparently has not been reported in larger animals. And there is little point in producing an unhealthy farm animal because only healthy productive farm animals will make the farmer any money.

b) *Cost*. Whereas the expense involved in making transgenic mice is relatively low, the cost of livestock is high. Most of the work to develop animal models of disease occurs in mice because much work involves gene knockout experiments and the stem cell lines needed for these experiments are currently only available for mice. TFA researchers, therefore, discard large animal embryonic constructs if they know that the constructs are analogous to transgenic mouse constructs that have led to problems in mice when brought to term. Farm animals are not used as models for specific gene function so many of the conditions that have been seen in mice are not likely to be seen in TFAs, and the number of at-risk TFAs is kept to a minimum.

c) *Frequency of attempts*. Government agencies are reticent to fund large animal transgenics since the 1985 USDA Beltsville growth hormone in pigs studies gave such bad phenotypes. They generally demand the mouse data mentioned in (1) if the large animal researchers do not first provide it. And private corporations are even less likely to take unnecessary financial risks than government.

A word of caution. Much transgenic animal research is now occurring in private industry labs where goats and sheep are being used to produce altered proteins in milk. Information about these animals is proprietary and not freely shared. Therefore, there may be incidents of transgenic animal suffering that only a few people know about. We simply don't know.
85 According to Baruch Brody, lecture at Michigan State University, 23 Oct 1995.

Conclusion

When social issues involve questions as momentous as global survival, ethicists should do more than analyze arguments and formulate principles. They should help us find stories to put our values into practice. In the essays collected here, I have tried to tell my story in the hope that others might find it instructive.

I once nearly believed that we should rule out the use of genetic engineering in agriculture. For reasons detailed in these pages, I changed my mind as I became convinced that intrinsic objections to ag biotech are weak, and that potentially adverse consequences can be effectively managed through government regulation. While continuing to have objections to the exploitation of vertebrates by our genetic engineers, I welcome the application of these scientists' considerable skills to plants.

I remain convinced that Wendell Berry offers a good story, and I now believe that ag biotech is compatible with it. We should learn to farm and eat well, in a way that respects the rights of sentient beings with futures, and does as little harm as possible. We need ethical principles and public policies to reform agriculture along humane and sustainable lines, so as to feed every child without exploiting people, land, or animals. We need institutions and practices that will allow us to leave farm ground in better shape than we found it, providing future generations in turn with the resources they will need to feed themselves.

No doubt, there will be a diversity of good stories to guide us, stories originating from specific places, enabling local inhabitants to farm in morally defensible ways, ways suited to their regions. Stories in which small and large farms may flourish, serving different needs. Stories in which people may seek the long-term common good of families and communities, while allowing individual stockholders in private corporations to profit from their investments. Stories in which animals, living alongside and among us, will provide us with various sources of nutrition, which we will take from them without compromising their welfare.

Perhaps, for example, some will find Wes Jackson's vision of a perennial polyculture the best story for those living in the high plains of the US. Perhaps, as Jackson himself has suggested, the tools of genetic engineering may be carefully integrated into this story so as to produce a perennial, genetically modified, high-yielding, grass variety native, more or less, to its place. A variety that can provide huge quantities of oil derived from its seeds while also helping to reduce the amount of acreage in production. So might ag biotech make good on its promise to help restore the beauty and integrity of the US's mid-region.

We need stories that show due consideration for animals. I have detailed my ongoing struggle with this issue. The animal rights view initially struck me as too strict, not capable of making relevant moral discriminations, and not allowing us to account for the fact that the mental lives of animals are not uniformly intense or complex. On further examination, however, that theory provides some supple and common-sense resources, such as the worse-off principle. On a reformist reading, the animal rights theory requires vegetarianism for those in developed countries, but also permits the use of many animals in research. Our challenge is to figure out how to do the least harm to humans and animals while involving humans and animals in a sustainable society. The practice of meat-eating among the world's well-off peoples does not accord with this story. Neither does the production of transgenic farm animals for food. Neither does the production of massive amounts of grain to be fed to animals for the production of meat. On the other hand, uses of animals in research that will demonstrably save human lives may on occasion be justifiable.

The production of transgenic plants is another story. It seems difficult to find persuasive grounds on which to object to the introduction of crops such as rice enhanced with vitamin A and iron. We have a duty to aid the needy; virus-free forms of cassava, millet, and wheat seem to be effective means of fulfilling this duty. Then again, there are good reasons of an environmental sort to object to crops grown in developed countries where the possibility exists for gene-flow to sexually compatible plants.

Coming full circle, we can apply these conclusions to the case with which we began. bGH is not, in the end, a technology we should endorse. It is important, however, to be clear about the reasons.

The rejection of bGH should have nothing to do with the fact that it is a product of genetic recombination. Nor that it will drive family farmers out of business. I object to bGH only on animal rights grounds; the technology represents a harm to dairy cows not justified by the gains to humans.

Anecdotal reports from dairy farmers suggest that bGH decreases the so-called useful lifetime of a cow by one milking cycle. Cows on bGH are put under additional stress by the use of the technology. The amount of added stress is not lethal to the animal, as the added stress is not life-threatening. However, the fact that farmers typically must slaughter bGH cows a year earlier than non-bGH cows suggests that the treated animal has a serious interest in not being injected with bGH. And the gains to poor consumers in lower milk prices achieved by bGH may be achieved by other means, such as taxes on the rich to provide free milk to the poor. Thus, we have strong moral reasons to think that bGH is not an ag biotechnology that fits with our humane sustainable story.

This negative judgment, however, must be set in the broader context of modern dairy production. Even dairy cows not on bGH are under tremendous physiological stress. They have been selected through traditional breeding methods for extraordinarily high productivity. Non-bGH production cows must eat continuously in order to produce the amount of milk their systems have been designed to produce.

Therefore, it is a mistake to focus on bGH. Our opposition, in truth, should not be directed at ag biotech but, rather, at the institution of modern dairy farming. We should oppose the practices of milking cows three times a day rather than twice a day; separating cows and calves at birth rather than allowing them to bond; allowing the birth of male calves to be slaughtered at a young age for veal; and killing cows when they are worn out from milking rather than letting them live out their natural life-spans.

A reformist animal rights ethic insists on a thorough reshaping of the industrial dairy farm. We should use technology to allow only female bovine fetuses to come to term. In this way, we will not need to slaughter veal calves. We should construct barnyards and pastures so that calves are allowed to form natural bonds with their mothers. In this way, we will preserve the natural instincts of the animals we are using while permitting them to satisfy the desires typical of their species. We should breed cows so that their capacity to produce milk does not threaten their long-term health. And, first and foremost, we should devise agronomic and economic husbandry systems that will allow cows to live out their natural life spans, not slaughtering them until, from the cow's perspective, the cow's life is no longer worth living.

Only in this broader reformist context does opposition to bGH assume its proper place in our story. The problem is not agricultural biotechnology. The problem is our current way of treating food animals.

bGH is not an anomaly. It is a symbol of the first wave of ag biotech products. I continue to agree with the global critics on this point, that the first wave of ag biotech products are premature and do not address the central problems of modern agriculture. However, the first wave may be justified on the grounds that it has provided private industry with revenue-enhancing products to pay for the research and development of a second wave of mature, ethically-sound, ag biotechnologies. The problematic first technologies may not be inseparable from a huge and expensive technocratic food system, an undemocratic social and cultural nexus controlled by a scientific and engineering elite unconcerned with the interests of most of the world's plain citizens and farmers. But the second wave may indeed be separable from MA, as the ability and resources to research, develop, and deploy the technology become more widely and democratically distributed.

There are reasons to continue to be skeptical that we can make the story suggested in the previous paragraphs come true. If ag biotech is going to make good on its champions' claims to feed the world's hungriest children, then several problems must be faced directly. Private corporations are putting their efforts, justifiably, into technologies likely to lead to increased profits. Herbicide resistance and Bt corn are good technologies for the companies that market them and the farmers that use them. But let us be honest. These technologies do not provide increased nutrient values in the staple crops of developing countries.

Moreover, developing countries face an uphill battle, in terms of legal, financial, and human resources, as they struggle to gain access to the products of ag biotech. Therefore, the challenge to those wishing to defend ag biotech on moral grounds is real. Corporations profiting from the first wave of technologies must make their patented technologies freely available to farmers and researchers in the poorest countries. They must share their legal, intellectual, and human resources with the neediest. They must recognize the contributions of past and present indigenous farmers who have helped to develop the germplasm now being manipulated in their labs.

Only in this way will local farmers and researchers be able to produce new GM strains of Africa's, Asia's, and Central America's basic crops. Only in this way will the proponents of ag biotech help us to progress beyond a donor mentality. Only in this way will we move toward a cooperative mentality, in which technologies are regarded not as gifts to be handed down, but as resources to be handed on. In a cooperative spirit, the poor become researchers and developers of their own, and eventually competitors with others.

By pursuing the right goals, we may indeed be able to compose together a story that encourages respect for human life, good farming, the feeding of children, and due consideration of nonhuman life. That story need not include Beltsville hogs with human genes, crops resistant to herbicides being grown only in blank perfections of fields to be fed to livestock for slaughter, or dairy cows whose lives are shortened by injections of stress-inducing proteins. That story should, however, include biotech products that will obviate the need to slaughter veal calves; that will reduce the price of food for our poorest consumers; that will make the most productive use of our arable acres so as to preserve lands for other flora and fauna; and that will help to meet the vitamin needs of the world's sickest children.

If we pursue these goals vigorously, collectively, and thoughtfully, ag biotech may eventually win over even its most confirmed critics.

Credits

Editors of the following journals have kindly granted permission to reprint modified versions of these articles:

"The Rights of Animals and Family Farmers," *Between the Species* 7 (Summer 1992): 153-156 [parts of which are found in the Introduction, ch. 3, and *passim*].

"The Case Against bGH," in *Agriculture and Human Values* 5 (Summer 1988): 36-52 [ch. 1].

"Genetically Engineered Herbicide Resistance, Part One," *Journal of Agricultural Ethics* 2 (1989): 263-306, and "Part Two," 3 (1990): 114-146 [ch. 2].

"The Costs and Benefits of bGH May Not Be Distributed Fairly," *Journal of Agricultural and Environmental Ethics* 4 (1991): 121-130 [ch. 5].

"Pigs and Piety: A Theocentric Perspective on Food Animals," *Between the Species* 8 (Summer 1992): 121-135 [ch. 3].

"Should We Genetically Engineer Hogs?" *Between the Species* 8 (Fall 1992): 196-202 [ch. 3].

"The Moral Irrelevance of Autonomy," *Between the Species* 8 (Winter 1992): 15-27 [ch. 3].

"Do Agriculturalists Need a New, an Ecocentric, Ethic?" *Agriculture and Human Values* 12 (Winter 1995): 2-16 [ch. 5].

"Is it Unnatural to Genetically Engineer Plants?" *Weed Science* 46 (1998): 647-651 [ch. 5].

"Research with Transgenic Animals: Obligations and Issues" *The Journal of BioLaw and Business* 2 (Autumn 1998): 51-54 [ch. 6].

Index

A

B

C

D

E

F

G

H

I

J

K

L

M

N

O

P

R

S

T

U

V

W